I0474336

THE CREATION SECRET

-Paul Garbett-

"*The Creation Secret*"
Version 2.2.2 (for Amazon)

© Paul Garbett 2012
(Original version © 2010)
All Rights Reserved
ISBN: 978-1475055221

Scripture taken from the HOLY BIBLE, NEW INTERNATIONAL VERSION ®. Copyright © 1973, 1978, 1984 by International Bible Society. Used by permission of Zondervan Publishing House. All rights reserved.

The "NIV" and "New International Version" trademarks are registered in the United States Patent and Trademark Office by International Bible Society. Use of either trademark requires the permission of International Bible Society.

Contents

Chapter 1 – Defining Science

Is Truth Knowable?

When most people today approach the subject of God, it is often assumed that such ideas can only ever be seen as a matter of opinion, and not of fact. It is generally perceived that the only certainty that can be acquired is from the personal experience of something supernatural – if such an experience were even possible. Whether it is Evolution or Creation, most people see the issue of the origin of man as something to be speculated about, and strong opinions need to be left at the door.

The real question however, is not about what each person thinks, but what *the truth* actually is. We need to face the fact that either God exists or he does not. We were either created by God, or we evolved by chance. Our opinion on the matter does not affect objective reality, yet our opinion has enormous implications on the way we live. The issue of Creation versus Evolution is contentious for this very reason. It is important to *resolve*, not just debate, because whatever is the foundation of our existence will shape our perception of the purpose and meaning of life itself, above every other issue.

There must be truth about God one way or another beyond our own perceptions, but can we really know that truth? Truth, by definition, is when an idea corresponds to reality. Something that we call 'true' should have supporting evidence in the reality it claims to describe, and therefore we ought to be able to confirm or refute ideas by the weight of that support. Things that are true can be shown to be true because they are consistent with all the observations we can make, and have an absence of conflicting facts. The more facts that are in agreement with an idea, the more confidence we can have in it being true.

We can find evidence supporting/refuting Creation/Evolution by examining the nature of the Universe, because both Creation and Evolution make claims about the nature of that Universe. For example, while Evolution suggests that reality is randomly generated, Creation claims that reality is an ordered invention. The reality that emerges from these alternative origins will reflect the nature of that origin for us to observe today. Therefore, by evaluating the nature of the Universe, we can discover the origin of the Universe. The more accurate we can be in that investigation, the more confident that we can be in the accuracy of our conclusions. Therefore, we must be careful to evaluate the facts honestly, and follow where they lead.

The Conflict Between 'Faith and 'Science'

Belief in God and belief in Science are often perceived as being incompatible. This most likely began with the dawning of what is often called the 'Modern Era', when Science gained widespread popularity and significance. It became increasingly common for there to be a competition between the 'religious' way of thinking and the 'scientific'. People began to see Science as explaining away many of the naive notions about the world that came from 'religion', and since we could understand the way things physically worked according to natural laws, we no longer needed the 'God explanation' to explain things we didn't understand. A huge amount of knowledge about the world has now been gained through scientific research, and it has become accepted as an authoritative method of finding the truth – paving the way for a secular (non-religious) society. Despite the fact that the scientific method was based on Christian principles,[1] many non-religious scientists have contributed much to our understanding of the world, and very early on it became accepted that one could be an effective scientist and not believe in an ultimate intelligence.[2] With an increasing number of intelligent people beginning to support 'Science' in opposition to religious views of the world, there has been a change in the perception of 'Faith', until today many people view it as only a matter of personal choice – having no real connection to the truth about the world we live in.

Modernism ushered in a conflict of views about the ultimate nature of the world we live in (Worldviews), and alternative explanations of reality (such as Evolution) have weakened the perceived cultural value of 'religion'. However, the emerging (Postmodern) culture we now live in has developed yet another mentality: it has come to embrace 'Religion' and 'Science' as *not conflicting* because they can be separated out into 'faith' and 'fact' respectively. That is, while 'religion' deals with what is morally and ethically acceptable (according to a particular perspective), 'Science' deals with facts about the real (physical) world. Science is about what we can know about reality, and 'faith' is becoming more about what we *cannot* know. Hence, many people now feel as though there is no value in contrasting 'Religion' and 'Science' because they inhabit different (non-overlapping) spheres in human understanding.

The (Modern Era) conflict between 'Faith' and 'Science' arose when these made conflicting claims about reality, centrally involving the Creation versus Evolution debate. The (Postmodern Era) separation of 'Faith' from

---

1 Don Batten and Jonathan Sarfati *15 Reasons to Take Genesis as History* (2006), pg 25.
2 It is necessary to assume an ultimate intelligence for Science to work. This is argued later in this chapter: The Universe (made of non-rational matter and energy) must be rationally constructed for a rational investigation to turn up anything. Hence an outside rational agent is a necessary precondition.

'Science' to avoid such conflict has resulted in a redefinition of the term 'Faith' so as to be unrelated to the facts of the world.[3] Hence, conflict between 'Religion' and 'Science' arises when both claim to be supported by facts (which are disputed), and the conflict is often avoided when 'religion' gives way to 'Science' and becomes *non-factual*. The overall result of these cultural trends has led to this time we now live in, when the Christian faith is considered by the majority to be intrinsically irrational. This is because it is seen to be based on ideas which have no firm grounding in reality, either because it claims no such foundation, or because that foundation has been refuted by 'modern Science'. Whilst this secular society has some respect for Christianity's ethical and moral precepts, these are not seen to have any absolute authority, because whilst they may agree with good conscience, they have no factual basis in reality.

The Nature of 'Science' and 'Facts'

Science is an enterprise of investigation and development which seeks to understand the truth about how the Universe functions, and therefore how new understanding and technology can be developed. The most central concept in Science is the relationship between *cause and effect*. The world we observe is a world of 'effects', and it is the scientist's job to logically infer what the most likely cause is for a related set of effects. Having understood this cause, and perhaps described it mathematically, the scientist can then use this understanding in different ways to generate new technology. Thus, through observation and investigation. the scientist can find and exploit the rationale built into the Universe.

Facts are what we observe. Conclusions or scientific theories are the best rationale that we can come up with in order to try to explain the cause of the facts. Scientific inquiry relies on *inductive reasoning* to connect cause(s) with effects. This is when a number of observations are made, and the idea is to assess those observations and logically connect them with the most reasonable cause that best explains all of the observed effects. The conclusions or theories that one makes are therefore dependent on the quality of the data that the scientist has to work from, and their ability to *think objectively and logically*.

Science is really about understanding the truth about reality, and as such, scientific conclusions should ideally be both *falsifiable* and *reproducible*. For instance, a mathematical formula is *reproducible* because it can be tested again and again with different numbers and proven to consistently give accurate results. Mathematical formula are also readily *falsifiable*, because in theory a bad formula would eventually be proven wrong by *counter examples*. In short, every valid conclusion one makes must have ample support with facts and must also be free of conflicting criticism.

---

3 This is explained more in Chapter 6.

This scientific/rational methodology is a systematic way to determine truth, and can be applied to many disciplines.

Conclusions we make based on things we can test and approve in this way have a much higher probability of being true than using other approaches.[4] There is also a huge difference between 'observational science' and 'historical science'. Observational science deals with what we can observe and test in the present. However, when it comes to the past, historical science employs a certain amount of guesswork, because it is impossible to observe and test events that have already happened. Whilst we can be objective about our observations in the present, it is impossible to do so when we cannot prove our conclusions by experiment, or potentially falsify those conclusions with evidence to the contrary. This being considered, theories about what has happened in the past can never be confirmed for their factuality. When we seek what has happened in history, we are restricted to only review all the facts we have available to us in *the present*, and then infer the *best explanation* or cause for what would account for these in the past.

Evolution & Naturalism

The term 'Evolution' is often used in a generic sense to encapsulate a number of concepts. Nevertheless, the general theory of Evolution (GTE) is the overall schema by which the world and everything in it came into existence *only by physical processes*. Evolutionary theory views the world as being strictly the result of time and random chance processes alone, and therefore Evolution really is the opposite of Creation, which claims that there is an external (supernatural) creative agent. There is not one fixed understanding of Evolution, rather, it is more realistically a collection of *Naturalistic* (without God/design) explanations of origins. Creative design is obviously not an acceptable explanation of the world to an evolutionist, because this is non-naturalistic, and therefore it is at odds with the whole concept of Evolution. Thus, Evolution is made up of explanations of the world which are consistent with *Naturalism*.

Naturalism is in fact a philosophical view of the world, not a science. It is the *belief* that the world is made only of what is natural or material/physical, and is obviously the opposite of Supernaturalism, which holds to the idea that part of reality is also immaterial. A natural origin for the Universe is the critical foundation of Naturalism, and if evolutionary theory is credible, Naturalism can be rationally accepted (not just an irrational philosophical belief). Naturalism also forms the essential basis for Atheism (the belief that no God exists) and other human-based belief systems.[5] Essentially, a successful evolutionary theory would give

---

4   For example, the Existential approach, where knowledge is gained through experience, which can be very subjective and tainted with bias.

5   Such as Humanism, Existentialism etc, which are based on there being no higher authority over

the factual basis for any set of beliefs which exclude God together with absolute moral values. Thus, evolutionists such as Richard Dawkins say that *"Darwin made it possible to be an intellectually fulfilled atheist"*[6] because he is viewed as providing that factual defence for Naturalism in the evolutionary theory.

These things considered, we can conclude that Evolution is unscientific, because, by definition, its proponents can only make conclusions that are consistent with the philosophy of Naturalism. Rather than an objective evaluation of the facts, Naturalism imposes a grid in which all the facts must fit in order to be acceptable to the evolutionist. Hence, objectivity is lost because an unchallenged *philosophical* criterion governs the validity of the conclusions one can make. Science may or may not support Evolution, but it would be dishonest to equate Evolution with a scientific approach, when it's very definition is biased against some facts and therefore, potentially, the truth.

Furthermore, since this unscientific bias towards Naturalism has become the predominant *cultural* norm, the possibility then exists for there to be a pool of scientific information supporting Theism (a belief in God) which has been excluded solely on the basis of its philosophical implications, and not its scientific merit. When all the facts we hear are being filtered through Naturalism, it is unlikely the public will hear any information in support of Theism. It could of course be argued that Evolution *is* scientific, and that it only appears to be biased towards Naturalism because there is simply no evidence that supports Theism. However, if Evolution was actually a scientific rendering of the facts, then it would be impossible to find information that compromises its integrity, and books such as this could not exist! The intent of this book is to overview this 'Creation Secret' to make people aware of the science that has been kept hidden because it conflicts with some people's philosophical biases.

God of the Gaps?

'God' is often seen as a non-scientific explanation, because it is argued, 'God' cannot be tested and validated in the laboratory.[7] 'God' is often seen as merely a cop-out explanation of things which we do not yet understand. The so-called 'God of the gaps' explanation is when we use the concept of 'God' to fill the gaps in human understanding in lieu of a better alternative. For example, lightning was once seen as having a supernatural origin. Furthermore, concluding that God is the cause of something is sometimes also seen to impede the progress of Science, because such an explanation prevents further investigation into an issue which might have a simple physical cause.

---

man's opinion and experience.
6   Richard Dawkins, *The Blind Watchmaker* (1986), pg 6.
7   Although, this still relies on a certain non-testable definition of God.

If we are to suppose that 'God' cannot be tested in the laboratory, and therefore cannot be a valid scientific explanation, then we would also have to discount anything that has happened in the past - including Evolution. If we based Science on only what we see happening in front of us, then obviously Evolution would never have become a theory! Evolution is a statement about what has happened in the past (an explanation of origin), but we do not see it happening in measurable time today.[8] The simple fact of the matter is that when we are speculating on things that happened in the past, we are unable to reproduce those causes, and therefore Science is relegated to choosing the *best explanation* that fits with what we can observe *today*. We cannot demonstrate either Creation or Evolution in the laboratory, because we cannot observe or replicate those past events. Instead, we must examine the 'fingerprints' of the world today and find the explanation which best fits the data we observe in the present.

God is often rejected as a scientific explanation because it is seen to only account for things that we do not yet understand. Some scientists argue that as we gain more knowledge about the world, we will no longer need to believe in God because those things we attribute to God will then be explained through Science. However, we could say this about many things already accepted by Science. For example, we could say that the theory of gravity is just something that we say because we do not understand why things keep falling to the ground! This of course is a silly argument, because the theory of gravity is our *best explanation* of the events we observe. Likewise, it is silly to argue that someone's acceptance of 'God' is because of lack of knowledge, if an intelligent creator best explains an intelligently ordered creation. For an increasing number of people, it is a deepening knowledge about the world that leads to a belief in a supernatural cause.

The truth has nothing to do with gaps in human understanding. The real problem is when one forces a wrong conclusion because of their inability to be objective and logical. Science will have faltered when conclusions presented are the result of someone's personal preference, philosophical bias, or potential financial gain. It would be wrong to believe that God best explains reality if that does not fit the facts, and likewise accept Evolution if it conflicts with observations. Science is about finding the truth, not proving our own viewpoints.

---

8   With the exception of 'micro-evolution'. This however is only a change within a species of plant or animal, and does not lead to large-scale change over millions of years (macro-evolution). See Chapter 2 of this book.

The Assumptions Of Science

Science is a rational inquiry of the Universe, and as such, it assumes a certain number of preconditions about the nature of the Universe for it to be successful. These preconditions are often just an assumption and are very seldom acknowledged. However, by explicitly stating these assumptions, we can have more clarity in saying what is and what isn't scientific. For instance, one precondition we assume is that the Universe is rationally ordered, so that when we study it there is something sensible to discover. Or put another way, we know from experience that the Universe (especially life) has an inherently high technology built into it. Therefore, when we study how it functions we can discover knowledge which is useful in building and improving our own technologies. For example, scientists have discovered that tiny flatworms called monogeneans secrete a 2-part super-glue that fixes them onto the gills and fins of other fish. If this glue can be copied by man, then scientists believe it could have an almost limitless potential as a commercial adhesive.[9]

Since Science has enjoyed so much success, we can know for certain that its founding assumptions do in fact represent the world that we live in. Four of those foundational concepts necessary for Science are as follows:

(1) There is a Cause for Every Effect – For everything that we observe to happen, we understand that there was a reason why it happened. Every physical event is preceded by another event which was sufficient to account for what we have observed happening. Matter is not capricious - things do not 'just happen' according to its own whim. Rather, matter follows a rationale where every event has an explanation to discover. We can make deductions about the nature of reality, and predict its next move, assuming that the same reasons why something happened in the past remain true for the future. Thus, matter follows a stream of causes and events, with each event leading to another in an orderly fashion.

Now, this series of events cannot continue on like this forever. Energy is usually lost in the process of change, and given an infinite amount of time all the energy of the Universe will disperse to the point that there would be no more potential for new events to occur.[10] If cause and effect hold true, we simply cannot have reached an infinite timespan of events in history, because we would have spent all the available energy. Thus, since energy (for new events) remains within the Universe, there must have been a definite beginning for the Universe in the finite past. It is also a

---

9 Flatworm Superglue, *Australian Geographic* **57**:120, January–March 2000. And Alexander Williams *God's Amazing Glue* www.creation.com/god-s-amazing-glue.

10 This situation is referred to as heat death. According to the second law of thermodynamics, at a certain point in the future all of the energy available to do 'work' will be spent, as the Universe attains a uniform energy distribution.

mathematical absurdity for the Universe to have been around forever. If the Universe was infinitely old, then today would be day=infinity. Each new day would be day=infinity+1. By the time you are reading this, it will be many days beyond infinity since it was written! This is absurd because infinity is a number that by definition cannot be exceeded, and so obviously we cannot have reached that time. Additionally, since the amount of available energy is always winding down, the further back we go in time the more concentrated the energy sources must have been, and the fewer original causes possible. The energy levels approaching the beginning would increase in order to account for all the energy and effects we have today. Therefore, when it comes to establishing a first cause, this would need to be the summation of all available energy in the Universe today (all-powerful/omnipotent). These are all logical certainties given that cause and effect, which is a necessary assumption of Science, holds true.

Additionally, since we know that matter remains in a state of rest until an external force is applied to it, then it follows that there must have been an external cause outside the physical Universe for there to have ever been any sort of beginning. Anything that begins to exist always owes its existence to something outside of itself. We can also surmise that since the Universe is made of matter and exists in time, the only thing that could cause this to exist would need to be beyond matter (super-natural) and time[11] (eternal). A Universe that starts without a cause would violate the whole concept of cause and effect – which Science is based on. Thus, the implications of accepting Science also means accepting a supernatural creation.

(2) The Universe Is Rationally Understandable – When we look at the world we expect it to follow a certain order and to make sense to us – which it does. We do not expect reality to be random or capricious, but we anticipate order and rationality in the way it is constructed and continues to operate. For example, whilst there are some things that appear to be random, like the placement of stars, the physics governing the movements of those stars is mathematically ordered. When we look at the world we fall into assuming that things are intelligently put together, and this gives a basis for using a rational investigation to learn from the Universe.

However, without the Universe being put together in an intelligent way at conception, there is no basis for assuming that there is anything rational or ordered to find in it. There is no reason why the Universe *ought* to be ordered or make sense to us, unless it was created by an entity that itself was ordered and intelligent. If there was no intelligent design in creation, then we would expect to find increasing *disorder* the deeper we look. We actually find just the opposite! The closer we look at the world, and in

---

11   Time is a function of matter. Without matter, time as we understand it does not exist, or at least does not move forward.

particular life, the more we see order and complexity at every level of reality. Matter is by definition not rational. It simply follows the continuous stream of cause and effect without freewill or intelligent input. It can therefore never self-order or become more cleverly constructed. In fact, according to the second law of thermodynamics, the opposite is true. That is, because of the limited amount of energy in the Universe, matter tends towards becoming more disordered as energy is dispersed and stabilised. Given that we can describe the interactions of matter with precise mathematical formulas, we can say for certain that matter is rationally constructed, yet matter is unmistakably non-rational, tending towards disorder all the time! When we find a non-rational entity rationally put together, the only reasonable explanation is that the rationale came from something outside itself. That is, intelligent ordering must have been programmed into creation at the beginning for it to exist today. It is necessary to assume the Universe makes sense for Science to have any point in existing, but such an assumption involves invoking and intelligent design at the point of creation. Affirming that we live in a Universe created by an infinitely intelligent being gives rationale to continuing scientific investigation, because the work of an infinitely creative being suggests that Science will be a virtually endless discovery of order and intelligence at every level. Hence, in Science there is an element of adventure and wonder as we progress.

(3) People Are Rational and Can Understand This Order – Human beings have the ability to think and reason, and this is what gives us ability to assess the world around us and make conclusions about it. Rationality by definition involves the use of the intellect to weigh the sensory information we receive, and make decisions on what to do with it. This rationality, or decision-making process, functions above and beyond the automated processing of sensory data. It involves the ability to have freewill/freethinking, because it is about *directing* one's ideas and *choosing* between ideas based on their *merit*. This is unique to the human mind as far as we know, because it seems that only we have the ability to interrupt cause and effect processing in order to make decisions about the relevance of information we are receiving through the senses. Although a computer can process information and follow instructions, it cannot think rationally because it cannot weigh the *merit* of that information or make its own decisions over its validity or truth. Thus, a computer is irrational and cannot understand the rationale of what it is processing. Nor can it make a mistake – it always does exactly what it is programmed to do, because there is no freewill to make wrong decisions.

Since all matter follows cause and effect in an automated fashion (irrational), nothing made of matter can be rational or have freewill.[12]

---

12 'Decision' involves a violation of cause and effect, because rather than everything happening as a result of the preceding event, 'decision' is when alternatives are assessed and selected based on an immaterial criteria (e.g. truth or error).

Now, if there was nothing more to people than the physical brain we can measure, then it would be impossible for us to have rationality. Being only a physical entity, the brain (like a computer) could only process information as an automated machine. To be rational, there must be an operator of that brain (or computer) that assesses the information and makes sensible decisions on what to do with it. Without such higher functioning there would be no such thing as 'scientific' or 'unscientific', because everything would just happen according to the chemistry (there would be no decision-making). Unlike a computer, we can *choose* to follow or ignore our inbuilt physical 'programming'. For example, we can choose to think about some things and not others, which demonstrates that people have some control over their brains. Since only a non-physical entity can be free from the constraints of cause and effect (computer-like automation), it makes a lot more sense to say that people are essentially non-physical or immaterial beings, inhabiting a physical body they govern. The ability to rationally assess information in a scientific manner can only be accounted for by explaining the human mind in this way as a combination of a rational (immaterial) person using a physical brain to interpret the physical world. We must assume rationality to be scientific, but to be scientific we must also accept the immaterial (spiritual) nature of man.

(4) We Can Use Logic to Infer Cause From Effect(s) – When a scientist attempts to explain the world by connecting effects with the corresponding causes, they use a *logic* called 'inductive reasoning'. This is about establishing what cause would sufficiently account for all the observed effects. Without logic, Science would have no basis to draw conclusions, but strangely we run into trouble when we attempt to define exactly what 'logic' is. Logic is not a human invention, but a human discovery.

There are principles we use to help make good decisions, such as Occam's Razor (which states that the simplest explanation is the preferred one). Occam's Razor *just happens* to be the way things are in reality, and if we don't follow it we often miss the best explanation of the facts. In other words, reality is generally governed by simple rules, and searching for the simplest explanation often aligns us with that. However, we did not *invent* Occam's razor - it is simply the truth about reality. This then is an example of a logic or *way of thinking* that we *ought* to follow. Now, being a 'way of thinking', logic is not a property of matter or energy, or anything else that we can measure in the world. It is a property of the human mind. If we discover logic as the right way a mind should think, then which mind is our reference point (the perfect thinking mind)? If Occam discovered Occam's Razor as a way of thinking, then where did he find it? Logic is a way of describing how the 'perfect mind' *ought to think*, and usually we discover it intrinsically or by trial and error. Consider however, that unless such a perfect mind does actually exist, then on what basis do we decide what is and is not logical? Anything could be called logical if logic was an arbitrary invention without external reference, and there would be no such thing as 'right' or 'wrong' logic. The scientific method is built around the concept that some deductions are true and others false, and hence an objective standard (the scientific method) for what is logical is assumed. Without an objective and fixed standard of logic the scientific method would be defeated, because anything could be 'scientific'. This assumption is built on the precondition of there being an objective reference point for the perfect mind that is intrinsically accessible to all scientists. Such a title can only be a reference to 'God', unless one of us has a perfect mind that everyone else can reference! Therefore, to believe in the scientific method, one must also subscribe to the sort of God that is present and perfect.

A Logical Worldview

Science does work, and since it could not work without assuming the above as facts, we have to conclude that these above statements are also fact. That is, cause and effect necessitate that there was a definite beginning of the Universe in the finite past, which was caused by a supernatural, rational entity. We really do live in a world where order and intelligence are built into creation, necessitating a rational, intelligent creator. The rationality in our own minds is due to the fact that we are spiritual beings operating a physical body. The governing laws of logic point to the accessibility of a perfect mind which could only be described as 'God'. When we look at the physical Universe, we expect it to make sense to us because we expect the rationality in our own minds to map onto the rationality that is imprinted into creation from a rational first cause.

These principles are seldom acknowledged, but they do in fact form part of the fundamental framework through which Science successfully operates. Therefore, we cannot use the scientific, or any other rational method, to attempt to refute a Worldview involving a supernatural creator! For instance, it makes no sense to claim that man is nothing more than a physical (irrational) entity, when we have used a rational method of investigation to conclude this!

Naturalism then cannot be sustained as a valid theory through the use of Science, because the basis of the scientific method is unmistakably supernaturalistic. The Worldview that allows Science to exist builds on ideas that prohibit ever reaching Naturalism by logic. You could use another method to argue against God perhaps, but no form of *logical* argumentation can be used to argue against the existence of logic! Therefore, it is unscientific to not believe in God, since a Theistic Worldview is a necessary precondition to scientific inquiry. In principle then, an Atheistic scientist is a contradiction in terms.[13] Perhaps this is the major reason why Science was largely founded by Theists. One author says that *"It is not until the time of Darwin that atheism appeared to accomplish anything significant in science."*[14], because until Darwin's Evolution, Science remained largely a Christian ministry.

Conclusion

Is It Intelligent To Believe In God?

It is intelligent to correctly connect the facts with the most likely explanation for those facts. Therefore, it is intelligent to believe in God if that's where the evidence leads. It is certainly not intelligent to ignore the plain and simple truth about the world just because it is not popular, or because it makes us feel uncomfortable. Science is about being objective in our search for truth, not allowing our biases to taint the quality of the conclusions we make. In many ways, it is necessary to assume an intelligent creator to begin to operate rationally in Science, and therefore Science can never *logically* arrive at a position of rejecting a Theistic Worldview or becoming a secular/Atheistic discipline. Since, then, there is no real conflict between God and Rational/Scientific discovery, the debate between 'religion' and 'Science' is merely an artefact of people's false definitions of what those are.

It is not intelligent to be a Science-minded person and yet not believe in God, because, as shown, Science itself is built on the foundation of Theism. Many times, people are simply not given the opportunity to develop an awareness of the truth, because the information they hear has already been filtered through a grid which removes any trace of evidence supporting Theism. Additionally, the issue is also made more complex because creation was an event that none of us were around to witness. However, using a scientific approach this book will show not only that Christian Theism is the best explanation of origins, but also that it is the only reasonable one in the light of the creation we study.

---

13 Many great scientists are non-Christians, because one can still believe in logic without believing in God. But this is inconsistent. One can love logical thinking without realising the ultimate origin of that.

14 Alex Williams The Biblical Origins of Science www.creation.com/the-biblical-origins-of-science-review-of-stark-for-the-glory-of-god#f2. Williams reviews *The Glory of God* by Rodney Stark, who reviewed the top 52 scientists that gave rise to the scientific revolution between 1543-1680, and found only 2 of them were *not* Christians. See also: www.creation.com/creation-scientists

Chapter 2 – Biological Evolution

What is Evolution?

It is often very difficult to define 'Evolution' because the term is used in a variety of ways. The general theory of Evolution (GTE) is the overall schema by which the Universe and everything in it developed naturally by itself (without any outside Creator). There is no one set mechanism of Evolution that has always been held as true, rather, evolutionary thinking is more realistically a collection of *Naturalistic explanations* on origins. Supernatural concepts like God and angels are excluded from evolutionary thinking by definition, because this would be to invoke a non-natural influence. Life is viewed as being merely the result of time and random chance processes, and therefore Evolution really is the opposite of Creation.

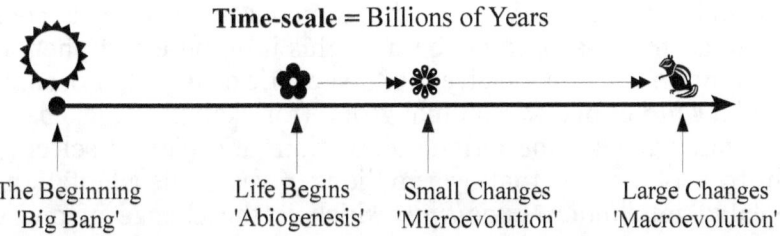

Based on a time-scale that stretches back about 15 billion years, the overall picture of Evolution includes 4 major concepts:

Within the framework of a very long time-scale:
(1) The 'Big Bang' – Cosmological Evolution is the idea that the Universe came into existence by itself, with everything we know being formed by chance.
(2) Abiogenesis – The origin of life was the spontaneous conversion of non-living chemicals into self-replicating, self-sustaining (living) systems.
(3) Microevolution – small genetic changes over time enable animals to adapt to different environments increasing their survival potential (Natural Selection).
(4) Macroevolution – over millions of years those small changes add up to large changes, and as a result new kinds of life develop.

This book will examine each of these concepts separately:

CH2 - Biological Evolution – Small genetic changes within a broad group of organisms (micro-evolution, Genetic variation, Speciation) is a well-known observation, and this represents the only aspect of 'Evolution' for which there is *experimental* evidence. However, in this current chapter we will examine the arguments for large-scale changes accounting for the origin of all species (macro-evolution).

CH3 - Abiogenesis – Change only occurs when there are successive generations of self-replicating organisms. In this chapter we will examine the likelihood of such systems originating by chance.

CH4 - The Big Bang – Without God the Universe would have to have been self-created. The scientific merit of this is examined in the context of the 'Big Bang'.

CH5 - The Age of the Earth – the evidence for the overall time-scale of Earth being billions of years old is examined.

Biological Evolution – Micro and Macro

No creationist has any difficulty with the concept of what is commonly referred to as 'micro-evolution', because this is an observed phenomenon. Micro-evolution is the name given to the genetic change that occurs within a particular *kind* of plant or animal group (for example 'chickens'), and is also known as genetic speciation, adaptation, and natural selection. It is thought by evolutionists that over millions of years this adaptation can be extrapolated into *macro-evolution,* which is the change of one type of animal or plant into another. By contrast, this has never been observed. Creationists do take issue with this, because it is non-scientific and woefully inadequate to explain the origin of all life.

MacroEvolution

Despite the fact there are many books and journals dedicated to giving factual support to Evolution, there are only about 5 basic arguments into which all those observations fit. When we clearly identify what the arguments for Evolution actually are, despite the fact that we can see there is much supporting evidence for each argument, the *arguments in themselves* are very weak. That is, there is a wealth of good science that goes into supporting statements which do not qualify as adequate arguments for Evolution (or against Creation). In fact, when isolated, we can see that those same arguments presented to support macro-evolution actually serve to support biblical Creationism! The 5 arguments commonly used in support of macro-evolution are:[15]

---

15   Several references in particular were used to source specific arguments: Wikipedia: http://en.wikipedia.org/wiki/Evidence_of_evolution, accessed 31/07/07. Berkeley University, http://evolution.berkeley.edu/evolibrary/article/lines_01, accessed 31/07/07. Otago University, Dr Ian Jamieson - University of Otago, BIOL 111 – Evolution Section, 2003. Curtis and Barnes, *Biology* Worth Publishers Inc 1989. However, this is common knowledge many people are familiar with.

(1) The Number of Species – when Darwin was formulating his ideas about *descent with modification*, the prevailing view at the time was that God created a limited number of species in their current locations, and that these species had a fixed anatomy. Darwin observed that there were far more diverse numbers of species in particular localities which could not be accounted for by the popular Creation model, and thus he saw the need for a theory that could explain the *diversity* of all living things. Evolution brought forward the idea that plants and animals can change to suit the environment, and that this accounted for the many changes within a particular kind.

☒ Darwin's evolutionary scenario was based on reforming an inadequate and *non-biblical* version of Creation. A more biblical Creation model affirms that a limited number of species survived Noah's flood (on-board the Ark), and through subsequent migration progressed to inhabit the whole world. According to the Bible, God did not create animals in their present location as fixed species. In fact, in Genesis plants and animals are created "according to their *kinds*", and the word kind is used throughout the books of Moses to mean a *broad group*. For example any 'kind' of hawk, heron, or great lizard was to be considered unclean (Lev 11), and this terminology recognizes some variation within the group. The extent of diversity seen in the Galapagos Islands, for example, was somewhat of a surprise to the world, but there is nothing about this that is inconsistent with the idea of a broad group of animals interbreeding and producing significant variation. Therefore, the variation observed in the natural world is not an argument against *biblical Creationism*, and perhaps if people at the time had held firmly to this way of thinking, there would have been no motivation to develop an evolutionary explanation for diversity.[16]

☒ Whilst Darwin's Natural Selection explained variation, it does nothing to account for the origin of that species in the first place. Intelligent creation and subsequent natural selection (biblical Creationism) is more consistent with the facts, because it suggests a large number of species was created by God with the capacity for variation to occur within that kind. Indeed, the diversity of species is a good argument *for* biblical Creation, because the very large numbers of different kinds of plants and animals, together with the large number of diverse features within these, is more consistent with an infinitely creative designer than random chances. It is not reasonable to suggest that such an incredible array of diversity across *every* species can be explained by breeding and simple copying errors alone. The Natural Selection Darwin observed only explains variation *within* a broad group, and is not an argument favouring Naturalism over Creation.

---

16 Darwin's Natural Selection provided a way of explaining variation, which the popular 'Progressive Creation' did not. Evolutionary theory then was developed in reaction to the flaws in a non-biblical version of Creation.

A diversity of nature suggests an infinite creative origin, and does more to bolster the case for biblical Creation rather than adding anything to evolutionary theory.

Biogeography – Darwin also noted that animals, as well as there being a diverse number of them, appear to be specifically suited to particular environments. The classic example was that of 'Darwin's finches', whose beaks were adapted to the particular types of food that are available in each region. Additionally, animals in particular are distributed around the world in ways that suggest they have evolved in those particular regions as a response to the particular environmental pressures present there. Therefore, the argument is that the geographical distribution of animals is more consistent with an evolving process of change, rather than the idea that they were created in their locations for a particular way of life by God.

☒ Again, the evolutionary model is reactionary to an unbiblical model of Creation. The Bible makes it clear that there was a worldwide flood that destroyed all animals and terrestrial environments about 4500 years ago (Gen 7:4). Thus, the population of plants and animals that we have today are a result of post-diluvian distribution (migration after the flood). Evolutionists claim the unity of all living things, and so must also come to terms with the fact that all animals have the same geographic source at some point in history (which they acknowledge[17]). Thus, both evolutionists and creationists argue a geographic origin of species, accept that in the case of biblical Creationism, animals are fully developed at this time (~2500 BC) and had the ability to migrate effectively to different parts of the globe. It is well-known that the physical isolation of a breeding group (genetic isolation) can quickly reach a point where it is impossible to interbreed with the parent population (if they were reunited). So, we would expect rapid changes to occur within broad groups at this time. For example, we might expect a number of different large cats (lions, tigers, leopards) to have developed from the one cat kind. Additionally, it is likely that many of these small populations did not find ideal living environments, resulting in certain populations becoming extinct, and therefore we would observe discontinuous populations in different parts of the world. What we observe today then is more consistent with creation, flood, migration, adaptation and extinction.

The diversity and geographical distribution of plants and animals has been put forward as an argument for the Evolution of species. However, these factors are readily accounted for by biblical Creationism, which provides an explanation more consistent with what we see today.

---

17 Page 9, http://en.wikipedia.org/wiki/Evidence_of_evolution, accessed 31/07/07.

(2) Imperfections – Since Evolution is an unguided process that occurs randomly through time by mutations etc, evolutionists expect there to be imperfections in design, as opposed to Creation, which ought to depict the genius and artistry of a super-intelligence. They argue that there are in fact many examples of imperfection in nature which are uncharacteristic of an intelligent designer. For example the inverted retina of the human eye and the panda's 'thumb', among others, are often cited as examples for which poor design is evident. Darwin noted that there were a variety of designs used in flowers such as orchids, some clearly better designed than others. Therefore, it made better sense to Darwin to view these as merely an artefact of imperfect, random modifications through time.

☒ Actually if Evolution was true, then we would expect nothing to be perfect. It would be hard to argue that perfection could ever be achieved through making random changes in DNA. DNA is more complex than the information on a music CD for example, and no one would argue that music could be perfected by introducing random changes! The term 'imperfection' implies that there is perfection everywhere to be seen except in a few particular cases. Unfortunately for evolutionists, these too are becoming rarer. It now appears that the more we learn about nature, the more we are discovering a higher wisdom in the design of living things. Many things that evolutionists once thought were leftover evolutionary waste, have now been found to play an important role after all. For example, it is now known that the inverted retina design in the human eye is necessary to keep us from going blind, because the connective tissue covering the retina actually protects it from oxidative damage caused by free radicals generated by UV light. This of course is not important at the bottom of the ocean, where we find the octopus and squid using a non-inverted retina.[18] The panda's thumb is also now considered to be a specialized feature which enables them to handle bamboo (their primary food source) with clever dexterity.[19]

☒ The orchids, which were argued as being examples of poor design, show ingenious design by mimicking the appearance of an insect, and tricking insects into pollinating itself. Some orchids are more ingeniously designed than others, but this is hardly an argument against any designer at all. The existence of cars of different designs to suit different driving situations does not refute that there was intelligence in their manufacture! A variety of simple designs amongst more elaborate ones is exactly what we might expect from an intelligent, artistic designer, who designed nature to communicate something of their character.

☒ The Bible only says that things were perfect at the beginning, and

---

18 Jonathan Sarfati *Refuting Evolution 2* (2002), chapter 7.
19 See *The Bamboozling Panda* @ creation.com and *Nature* 397(6717): pg 309-310.

suggests that this perfection is being corrupted over time. What we are looking at today is a copy of a copy, and we can expect there to be a growing number of flaws developing due to accumulated mutations.

Vestigial Structures - If it was true that plants and animals change from one form into another (macro-evolution), then in the process of change we could expect to see the imperfections or leftover parts from previous ancestors. The argument from evolutionists is that structures that seem to serve no purpose are leftover parts from previous species, and are therefore strong support for macro-evolution.

☒ Scadding, an *evolutionary* Zoologist, said that *"Vestigial organs provide no evidence for evolutionary theory."*[20] This is because it is impossible to prove that something has no function, and we simply may not understand what each device is used for. As we have learned more about these, the number of vestigial parts has been reduced from hundreds to a few doubtful cases, and it would be safe to say that these could be dismissed now. Examples like the whale's and snake's 'rudimentary hind limbs' (essential for mating), the human tail bone (an anchor point for essential muscles), and the human appendix (an important role in maintaining intestinal health) have all been used in evolutionary textbooks, but are now accepted as functioning anatomical features.[21]

Developmental Biology - this was once thought to also show the likelihood of Evolution, because in the embryonic stage many organisms look extremely similar. It was once argued by evolutionists that developing embryos displayed characteristics of their evolutionary ancestors. For example, the developing human embryo appeared to have gill slits not unlike a fish.

☒ We now know that it is the DNA that determines the final anatomy of any organism. All animals begin from a single cell, and hence look more similar the closer to that stage they are. However, it is well-known that from conception an organism's DNA is complete, and therefore the appearance of early embryos has nothing to do with their final anatomy or "recapitulating" earlier evolutionary stages (evolutionists did not know about DNA when this argument was formed). 'Gill slits' on humans are the initial stages of development of the thyroid gland – which has a function quite unlike fish gills.

In reality, evidence to support the argument of imperfections/vestigial structures does not draw on things that we find serious fault with, rather

---

20 S.R. Scadding, 'Do vestigial organs provide evidence for evolution?' Evolutionary Theory 5:173-176, 1981.
21 See 'Vestigial' Organs Questions and Answers www.creation.com/vestigial-organs-questions-and-answers.

evolutionists give examples of things for which we *don't yet understand* their function ('Evolution of the gaps' argument). Hence, they are not arguments against a supernatural creation, but unanswered questions about the *nature* of creation. They do not disprove that God created something, we just don't understand why something appears as it is. This is in fact just what we would expect to find if creation was made by a designer with a higher intelligence and creativity than ourselves.

(3) MicroEvolution – It has been observed that organisms have enormous variation potential, and can undergo fairly rapid adaptation and speciation to increase their survival probability over successive generations. For example, we have observed artificial selection of species such as dogs, chickens and horses, and natural selection (due to environmental pressures like temperature and lack of food) of species such as Darwin's finches. There are many examples of variations that produce different varieties of a particular broad category of organism. We have also observed changes which produce an altered version of a particular species due to mutations. It is also possible for organisms to remain the same anatomically yet seemingly develop alternate biochemical pathways to adopt poison resistance (e.g. myxomatosis resistant rabbits), drug resistance (e.g. vancomycin resistant Staphylococcus aureus), and pesticide resistance (e.g. DDT resistant mosquitoes).

The study of Genetics now explains the variation within successive generations occurring as a result, mostly, of sexual recombination of chromosomes[22], but can also involve mutations (and gene splicing (swapping) in bacteria). In the day of Darwin however, Genetics was still an undiscovered science, and it was not known that this variability was a result of genetic programming in the DNA. The physical structure of an organism cannot change without there being an internal genetic change. This means that an organism cannot adapt in response to environmental pressures (such as longer fur in colder climates), but rather, the genetics for particular traits of varying advantages and disadvantages already exist in the DNA. Natural selection is the process of *eliminating* those organisms which are *unfit* for survival, or those that are unable to compete with the more suited varieties. As a result of genetic analysis of such changes, no scientist would argue that new information is added into the gene pool through recombination of genes, because selection processes only select against the unfit genes already present. However, it is still thought by evolutionists that these small genetic changes (micro-evolution) that occur in populations, together with mutations, can add up to large changes over millions of years, eventually producing entirely new species (macro-evolution). Therefore, this 'descent with modification' is the origin of *every* species.

---

22 The offspring only inherits half of the parent's DNA (the parent having 2 copies), and this produces variation from the parents.

☒ Recombination – Animals that breed sexually (mammals for example) donate only one of the two copies of their genetic information to the next generation, and this is combined with one set from the other parent - making a complete set. For example, the genetic code in humans comprises 46 chromosomes, or segments of DNA, and 23 of these are from each parent. Variation occurs in the offspring according to which chromosomes in particular were transferred from the parents.[23] As an example, if a cat with long fur breeds with a cat with short fur, then there can be a range of different fur lengths found in the offspring. If someone was to take only those animals which had long fur and continue to breed from them, it would eventually result in a population that develops exclusively only that type of fur length. This selection isolates the population genetically, and for the sake of emphasizing one particular trait, there is a loss of variability (the ability to produce short fur). Strawberries used to be half the size they are now, but they also tasted twice as good. The better looking strawberries sell better, and have been selected by farmers. Over time however, continual selection for bigger strawberries results in that particular trait being emphasized at the sacrifice of the trait for rich taste. The important thing to note however, is that these changes always occur *within* clearly defined genetic boundaries. The common ancestor of all the different breeds of dogs was still a dog! No one has ever seen a dog produce a non-dog variety by recombination of existing genes – nor is it considered possible through this process alone.

Sexual recombination is well-known (in the science of Genetics) to be only a reshuffling of those genes which are already present, and therefore does not introduce anything new into the population. If we extrapolate this process over millions of years we do not, however, arrive at a macro-evolutionary change (an entirely different species). As we look back in time, we eventually arrive at the 'wild type' of a species, which had the original genetics to code for the variability which produced all the subsequent generations we now see. For example, we can suppose that in the past there must have been a kind of 'super dog' to account for all the different breeds we have today.[24] Since this process, genetically speaking, is going from more complex genetics (high programmed adaptability) to simple or refined genetics (only a few specific traits), this in fact is an example of *devolution*. That is, the observed changes we see in micro-evolution is in the *opposite direction of simple to complex* (molecules-to-man changes), and thus does not support the general theory of Evolution. In fact, it is evidence against it.

The ability to adapt to different environments over successive

---

23 Chromosomes come in pairs, and only one is selected to make a gamete. Each couple of humans can make $2^{46} = 70,368,744,177,664$ different children.
24 In fact, we know this to be the case, because all the different breeds of domestic dogs, horses, cows, cats and chickens were all bred by man.

generations is evidence for clever design rather than blind Evolution. We would say that a car which could automatically adjust its suspension according to the road conditions more advanced than one where that was fixed. Sexual recombination provides variability and adaptation potential in the long-term without compromising the genetic code, and this *programmed variability* supports design more than macro-evolution. The adaptability of plants and animals is more consistent with forward thinking and consideration of the long-term survivability of a species from a designer wishing to "fill the Earth" with life.

☒ Mutations – Mutations are very common, and for this reason the recombination of genes that occurs through sexual reproduction is very advantageous.[25] There are also a number of micro-molecular machines that make error corrections to the DNA to prevent mutations during replication. It would be very odd for these to be produced by Evolution, since these mechanisms appear to be anti-evolution devices! However, some mutations do occur on the level of the DNA coding which sneak through and affect the overall biochemistry or anatomy of a creature. For example, sickle cell anaemia and phenlyketonuria are biochemical diseases which are caused in humans by single-point mutations in the DNA. Sometimes single-point mutations can be an advantage - as in the case of blind cave fish.[26] Even in such cases, these *deleterious* changes are always a *loss* of genetic information. The cave fish have lost their eyesight, and that happens to advantage them in a dark environment. Since mutations are by definition a corruption of existing information, they are an example of *devolution* - in the same way that deleting words randomly out of an encyclopaedia is the opposite of writing a book. The loss of genetic information through mutation is the opposite of molecules-to-man Evolution (GTE).

The real question, then, is whether the information present in the genes of any given animal is like an encyclopaedia, or merely a random sequence. If one believes that all the information already present is the result of random changes, then it is not perfect, and therefore it might be possible to make some changes without there being adverse affects (Evolution). The consistent observations of Science, however, have shown that in each case presented it appears that genetic information is almost perfect, and even point mutations (one single letter difference from the parent) can result in catastrophic genetic failure (death). The examples given earlier of phenyketonuria and sickle cell anaemia are the result of only one letter mis-copied out of more than 3 billion base pairs in the human genome. These demonstrate the *frailty* of the

---

25 Because we have 2 copies of every gene, if something does happen to go wrong, there is a backup.
26 Since they live in dark caves, eyes are useless. Fish born without eyes have an advantage of their eyes not being damaged by swimming into the walls. This is of course a loss in the information needed to code for eyes, and an example of *devolution*.

human genetic code, and hence how intricately designed human life must be. Biochemical machinery (enzymes etc) in fact requires very precise genetic coding to form the specific 3-dimensional construction necessary to control chemical reactions, and it is not reasonable to believe that these are the result of tinkering with the random letters. If it really was the other way around, then why don't more people live near areas of high nuclear radiation, because this increases the likelihood of mutations! Common sense affirms mutations as destructive, not adding innovative changes. Another example is that of the ever-increasing number of genetic diseases in the human population. Whether people were created or evolved, obviously the original human population could not have carried all these diseases! Genetic mistakes are accumulating in the human population all the time. This is more consistent with the idea of people having been created with a perfect genetic code, which is being corrupted over time (biblical Creation), than for people to have started with a genetic code resulting from random changes that over time is being perfected (Evolution).

Observation and common sense tell us that longer time periods are no friend to Evolution, because the longer something is allowed to mutate, the more likely it is that it will incur a lethal mutation. The DNA only needs to incur one mutation at the wrong place on the genome and the organism is dead. This is exceedingly *more likely* than there being an advantageous, complete and functioning set of mutations forming, for example, a new functioning protein. Take for example the idea of a rat evolving into a bat: this would require a very large number of *specific and simultaneous* changes to different parts of the DNA. For there would have to be the development of new bones, new skin, new muscles etc, without there *ever being one wrong mutation*. It is obvious that no *interdependent* biological or biochemical system could be the result of mutation, because random changes cannot simultaneously invent new codes in different localities without mistakes – and this is necessary for organisms to evolve. There would need to be thousands upon thousands of changes in the right place without ever having one in the wrong place - an impossible demand for random changes to achieve. Recombination does not cross genetic boundaries within a kind, and therefore mutation alone is a totally inadequate explanation of the origin of new functioning organisms. Moreover, the changes we see though mutations do more to reinforce a creationist position. Given that mutations are occurring all the time, only serve to corrupt things already present, and bad mutations can kill, then it is very likely that the varieties of organisms that exist today are descended from genetically more perfect examples. Mutations are an example of *devolution*, because the change that occurs is never going to increase complexity or generate new coded information.

☒ Biochemistry – Micro-organisms are often given as examples of rapidly evolving species. For example, some bacteria are known to rapidly develop drug resistance. It is also observed that these 'super-bugs' are dependent on very clean conditions to survive. We now know that this is because they are threatened by the 'wild-type' of bacteria, which are in fact stronger than the so called super-bugs, and overwhelm them when competing for limited resources. As it turns out, super-bugs are actually not as super as first thought, because in becoming as specialized as they are, they have in fact lost some of their survivability. Whilst they might thrive in an otherwise sterile environment, they are often too weak to survive elsewhere. It has also been said that the way to be cured from a super-bug is to swim in the sea and wallow in the mud, because in doing so you will pick up enough normal bacteria to drive out the original infection!

These bugs are not evolving at all, but rather, the selection pressure of the environment that we create for them only eliminates the normal bacteria, therefore amplifying the resistant strain – but this resistance was already present in the original population. Vials containing penicillin-resistant bacteria have even been found that pre-date the invention of penicillin! What has happened to the bug on a genetic level is that the gene controlling the manufacture of the enzyme *penicillinase* (which breaks down the penicillin drug) has mutated and broken. The resistant bacteria cell produces far too much of this enzyme, giving it unusual super-strength against the drug. However, with the regulatory gene switch jammed on like this, the organism wastes too many of its resources producing *penicillinase,* and ends up being weaker in other areas because of it. What we see happening here is an example of a mutation damaging an existing part of the organism's DNA, which coincidentally gives it an advantage under certain conditions. This is a deleterious mutation of something that already exists in the organism's genome, and is again an example of *devolution.*

Some micro-organisms are also able to increase their adaptability by exchanging genetic information. This process again is more consistent with design, because what we are seeing in this instance is a clever process by which these organisms can increase the probability of their long-term survival. Again, these changes do not support macro-evolution, because such a process is only dealing with already existing genetic information, and there is no mechanism to generate that information to begin with, nor to add anything innovative to it.

Therefore, the small changes we see happening that are referred to as micro-evolution *do not* add up to the large changes necessary for macro-evolution. The most important reason why this is the case is because the changes that occur in every instance are in the *wrong direction* – leading from complex to simple, not simple to complex as

23

needed to support Evolution. Programmed adaptability through recombination of chromosomes or gene swapping (in micro-organisms) involves a reshuffling of existing genetic information. Random changes made to the DNA through mutation remove genetic information. Some have no effect (eukaryotes have a backup set of DNA), many are fatal, and in only a few rare cases the genetic disadvantage turns out to be advantageous. This *devolution* implies that organisms are getting genetically worse (as evidenced by increasing rates of genetic disease). It is in fact more reasonable to suggest that plants and animals have descended from a fewer number of genetically superior (perhaps perfect) species. Thus, all types of micro-evolution are strong support for biblical Creation, and do much to discredit Evolution.

In 1980 many of the world's leading experts in Evolution gathered for a conference in Chicago:

> *"The central question of the Chicago conference was whether the mechanisms underlying micro-evolution can be extrapolated to explain the phenomena of macro-evolution. At the risk of doing violence to the positions of some of the people at the meeting, the answer can be given as a clear, No."*[27]

(4) Homology – The arguments of *diversity* and *homology* are often placed alongside one another in evolutionary textbooks and the like. We have already looked at how the diversity (differences) of living things is perceived by evolutionists to be evidence for Evolution. Now we see that the opposite argument is also used as evidence for Evolution. That is, the similarity (homology) that we find amongst living things is hailed as evidence for Evolution. Homologous anatomical features, such as the similar forelimbs of Tetrapods, are featured in many Evolutionary texts. On the cellular level, it is true that *all* organisms have basically the same ultrastructure and biochemical machinery. The argument is that these homologies demonstrate that all living things are essentially descended through time from one original source - the original primordial life-form. The suggestion is that '*descent with modification*' (macro-evolution) best explains this phenomenon.[28]

☒ However, similarities (homologies) that unite life do seem to be a stronger argument for macro-evolution than diversity, because random chances seem woefully inadequate to explain the millions of clever designs we see in the living world. Whereas, it does make some sense that random changes to an original organism could leave an essence of unity in all life. Nevertheless, similarities are also a very weak argument

---

27 Roger Lewin, *Evolutionary Theory Under Fire*, Science 210:883, 21 November 1980. Quoted in *One Blood* – Ken Ham, Dr Don Batten and Dr Carl Weiland @ creation.com
28 Page 5 Berkeley University, http://evolution.berkeley.edu/evolibrary/article/lines_01, accessed 31/07/07

when contrasted with design. If evolutionists are going to argue that the differences in living things are evidence for Evolution (diversity - section one above), then it is highly suspicious that they also argue that the similarities are evidence for Evolution as well! Descent with modification is one explanation, but it is *certainly not the best one.* Unity only suggests a common origin, and therefore it can be used as a good argument for Creation! Despite the remarkable diversity life demonstrates, the unity of all life suggests a single origin. If every creature was different and there was no unity in nature, evolutionists would use this as evidence to say that creatures have evolved randomly at every level, and hence everything is different now that the original parent species have died out. If every creature was essentially the same with only minor changes, then again, evolutionists would regard this as evidence of a random drift away from an original organism. What we do in fact see however, is that life is both *similar and distinctive, and this makes it* impossible to envisage either of these naturalistic scenarios!

| Diversity Only | Unity Only | Unity and Diversity (Observed) |

For example, whilst we could argue that a whale has features that are similar to a fish, it could not possibly have evolved from a fish because of the overwhelming differences! (Evolutionists argue they evolved from land mammals) The reality is, that the fish-like similarities of the whale *have nothing to do with descent* with modification from the fish.[29] Such similar features are considered by evolutionists to have arrived by change through '*convergent evolution*', and do not indicate ancestry. The octopus's eye is another example: considered by many to be very similar to our own eye, but because the octopus is not 'closely related', it is considered to have evolved independently. There are several kinds of animals capable of flight: insects such as bees and butterflies, birds, pterodactyls (extinct) and bats. Even despite their similar design features, no one argues these are closely related (in evolutionary terms). This, then, is another example where common features have nothing to do with descent with modification. Evolutionists themselves know that homologies may or may not indicate ancestry, and they are only counted as evidence for Evolution if they already fit into the evolutionary tree they have constructed. In

---

29 The Evolutionary argument for this is that the ability to swim evolved independently (convergent evolution) when mammals took to the oceans. Therefore, the swimming features whales have are unrelated to fish.

other words, such *evidence* may be discarded if it doesn't fit with the theory of Evolution.

The common ancestry explanation looks particularly weak when we try to use Evolution to explain mosaic species, such as the platypus, which has a number of homologous features, all of which are thought to have evolved independently. Nobody regards the platypus as having evolved into (the common ancestor for) everything that it shares features with (ducks, mammals, beavers, reptiles, fish etc.), neither is anyone bold enough to suggest it evolved from a number of distinctive species. Therefore, because it has not evolved *into* a number of animals, or *from* a number of animals, evolutionists suggest its many homologous structures randomly evolved to coincidentally look like other animals. Thus, its duck-like bill has no connection with ducks, and just happens to be the same! This is an admission that homologies really have nothing to do with descent from one organism to another, and that the evolutionary tree is constructed despite the evidence to the contrary. On the other hand, shared homologous features fits perfectly with design in unison with diverse features, because they show unity of origin in a diverse variety of organisms.

If God wanted creation to look *unlike* Evolution, it would appear exactly as it is, with an almost infinite diversity interwoven with an almost infinite unity. Both of these observations when put together make it impossible to construct a model of descent to explain common origin, and instead point to a single infinite, intelligent source.

DNA – this is rapidly becoming considered to be a powerful argument for common descent, because it shows similarity at the genetic level.

☒ Similarity still only shows a *common source* and not common descent, and needs to be considered alongside diversity. When we are consistent with both evidences, an intelligent creation which is designed to look unlike Evolution is more reasonable. Consider that there are DNA similarities between organisms that are not considered to be closely related (in evolutionary terms). For example, sharks and camels have been shown to possess the same unusual single chained structure in one of their immune proteins.[30] Again, this genetic similarity is dismissed because a close relationship between these two species is not consistent with Evolution. *"We also share about 50% of our DNA with bananas, and that doesn't make us half bananas, either from the waist up or the waist down."*[31]

Whether it is similarities in DNA or anatomy, seemingly non-functioning parts or imperfect structures, none of these are evidence for macro-

---

30 New Scientist, 160(2154): pg 23 (1998).
31 Steve Johns quoted in article *Furry Little Humans* @ creation.com.

evolution. At best, they are merely the minor aspects of creation that we don't understand, when everything else appears to be intelligently and exquisitely designed. At worst, they provide strong evidence for Creation, because they demonstrate a rich unity and integrity in nature consistent with a single source, while at the same time defying descent theories. Observations are more consistent with the idea of an infinitely creative (generates huge diversity) single source (generates huge similarity).

(5) The Fossil Record – Fossils are the petrified remains of plants and animals which once lived, but are now entombed in rock. There are billions of fossils all over the Earth, revealing an important record of past biology. One of the predictions of evolutionary theory is that there will be some record of the development of species through time evident in the fossil record. As more of the fossil record has been exposed since the time of Darwin, evolutionists claim that this prediction is being confirmed. They claim that it has been observed that simpler forms of organisms generally precede and pre-date the more complex ones through time. For example, they argue that fossilized horses show a general progression through time indicating gradual changes. Evolutionists also claim that there are examples of 'transition fossils' of animals regarded as being intermediate between species, e.g. Pakicetus (land animal-whale) and archaeopteryx (reptile-bird). It is argued that some 'missing links' and truncated progressions are expected because of the fragmentary and imperfect nature of the fossil record, and therefore these are not considered as evidence against the general trends.

☒ Darwin himself said that the lack of fossils supporting his theory was *"perhaps the most obvious and serious objection"* to his theory. David Raup, the curator of the Field Museum of Natural History (Chicago) in 1979 said: *"We are now one hundred and twenty years after Darwin and the knowledge of the fossil record has been greatly expanded ... but the situation hasn't changed much. We have even fewer examples of evolutionary transition than we had in Darwin's time."*[32]

☒ An important thing to note is that the sequence of fossils is determined *by the assumption* of Evolution. The rock layers are *assumed* by evolutionary geologists to have been laid down over millions of years, and the general sequence that they are arranged into is according to the *indicator fossils* which are found within them.[33] Therefore the oldest rock layers are inferred as such by the appearance of simple and small fossils, with the more complex and 'highly evolved' ones representing the youngest rocks on top. Since this sequence is therefore based *on the assumption* of Evolution from simple to complex over time, then it is

---

32 David Raup, *Conflicts Between Darwin and Paleontology*, Bulletin, Field Museum of Natural History (1979).

33 The indicator fossils are placed in sequence according to the evolutionary tree. Generally, this follows descent with modification, and simple to complex.

unscientific to say that the fossils represent a progressing history from simple to complex – they are arranged this way by evolutionists!

☒ The fossil record is only consistent with Evolution if we assume millions of years of geological time. If the geological layers were laid down in huge volumes in a short time, then the progression from bottom to top of simple (small) to complex (big) does not represent Evolution over billions of years. Alternatively, using a flood mechanism to generate those fossils (see chapter 5), the progression of small to large species could be explained by hydrodynamic sorting processes that would occur under water. This is a simple and an obvious explanation, especially given that almost all fossils are found in *sedimentary rock* - which is formed under water by definition. This conclusion is avoided because it is *assumed* that those geological layers were laid down over millions of years. However, we now know that many of the so-called 'simple' organisms in the lowest rocks are not simple at all, but are every bit as sophisticated as 'higher' organisms (they are just smaller). For example, Trilobites, which are thought by evolutionists to be one of the earliest multi-celled species to have existed, had schizochroal eyes made of many lenses. These are able to correct for spherical aberration giving an undistorted image under water. They are thought by some to be *"the most sophisticated optical systems ever utilized by any organism".*[34] There are many more examples of organisms being claimed to be simple for the sake of the perceived 'evolutionary sequence', but in reality being every bit as complex as animals supposedly millions of years more advanced. The fossil record does not show simple to complex over time, it shows small to large over depth. Even the wide-spread prevalence of so many well-preserved fossils all over the world (used as evidence for Evolution) is more consistent with the idea of a large-scale catastrophe.[35] Fossilization is something we do not see happening today (at least on the same scale), and so we ought to assume radically different circumstances from today occurred to account for them. Rather than representing a progression through time, the layers of rock and the fossils in them could be stratified and sorted respectively according to hydrodynamic processes underwater, rather than Evolution.[36]

This scenario is a better explanation than the notion that successive waves of water (caused by uplift and subsidence of continental shelves) in turn fossilized the species we find in them. Slow and gradual processes are incompatible with the abundance of terrestrial mammals

---

34 Andrew Snelling, *In Six Days* pg 274.
35 See Chapter 5 (of this book). Exceptional circumstances would be required to account for the billions of fossils we find all around the world. Such fossils in sedimentary rocks layers are not forming today.
36 Rock layers are known to self-stratify under moving waters. For example see *Spontaneous Stratification of Granular Mixtures,* Nature **386**, 379-382 (27 March 1997), and *Sedimentation Experiments: Nature finally catches up!* http://creation.com/content/view/1541

buried with fish and compressed under a mountain of ocean sediment. It is far more sensible to conclude that there was a large-scale event which deposited plants and animals according to hydrodynamics and not age. In fact, this is the only plausible explanation for the abundance of any such fossils alongside massive deposits of coal and oil etc.

☒ Abrupt Appearance - Overwhelmingly, the fossil record shows *abrupt appearance* followed by *stasis*. "*The record shows that new biological designs typically make abrupt appearances, followed by non-evolution, called stasis.*"[37] This observation has led some evolutionists to conclude that Evolution is mostly static, and then proceeds in short bursts which are too rapid to be captured by the fossilization process (called Punctuated Equilibrium[38]). 'Pre-Cambrian' rock is generally unstratified (not sedimentary) and does not contain many fossils, whereas there are a multitude of fossils located in the sedimentary rock layers occurring just above that. Almost every kind of animal appears abruptly above the pre-Cambrian strata seemingly without a gradual lineage of development proceding it (called the 'Cambrian explosion'). Evolution cannot progress in anything other than random incremental changes, and this should leave a rich history of development. However, it actually appears as though fossilization of major species didn't even occur until a certain point in history, then it occurred everywhere all at once. This is more consistent with the idea that animals and plants were created in a short time period, and were mostly killed in one large-scale disaster which also encased them in the sedimentary rock we see today (biblical Creationism).

☒ Stasis – the second major observation of stasis also has significant implications. It is easy enough to identify what a fossilized horse, for example, looks like because they are very *similar to the ones we find alive today!* This is not what we would expect from Evolution, because if Evolution is defined as an undirected process of random changes and selection, we do not expect to find any organisms which have remained unchanged (un-evolved) for millions of years! The fossil record is the record of the past, and if many organisms appear no different from the ones we have today, then this surely is evidence of *no evolution*. The common argument against this is that once a high survivability has been reached, it would be expected that an organism would remain static (variations from this optimum would be eliminated). If this was true however, then there would certainly not be the diversity of organisms we see today. With the variety that we do have, one arguing for Evolution would have to argue that it must have progressed beyond organisms which were very well suited for survival. No change equals no evolution – and the fossils so clearly show this 'stasis' (non-

---

37 Walter James ReMine *The Biotic Message* pg 325.
38 A theory formulated by evolutionists Stephen Gould and Niles Eldredge, see article *Punctuated Equilibrium: come of age?* @ creation.com

evolution) that it is often included as part of the theory of Evolution!

A fossil record tree diagram not unlike the one pictured here is commonly found in *evolutionary material,* and is a frank admission of the lack of direct evidence for Evolution in the fossils. They show time on the vertical axis, and the number of fossilized species discovered on the horizontal axis. Thus, vertical lines represent no change or non-evolution, and the more curved or horizontal the lines are the more evolutionary change has been observed. Curved connectors are typically added to these diagrams to represent ancestral links, or where one animal has changed into another. Notice here that the only curved and

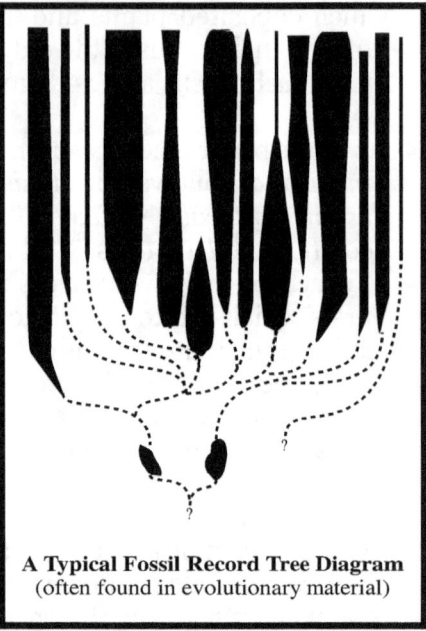

A Typical Fossil Record Tree Diagram
(often found in evolutionary material)

horizontal lines on the diagram are those that are dashed connectors. In other words, the evolutionary change is entirely a product of imagination! What the evidence presented in evolutionary textbooks and the like actually show is that there is a sudden appearance of fully formed organisms, no change over time, and no transition from one organism into another. Notice also, such diagrams often depict lines leading to unknown ancestors, unknown origin of life, and totally unknown lineage for certain species (far right[39]).

☒ It is also true that we can find a larger number of species of plants and animals in the fossil record than are alive today (because of extinction – dinosaurs for example). This is consistent with the idea that there were a certain number of created kinds, and when some of those become extinct, the overall number of organisms is reduced. Whereas, according to the evolutionary scenario, overall the total number of species would need to increase over time.

What the fossil record does not show is one animal changing into another. We are only able to dig up species that have been fossilized, but not a complete record of what has happened in the past. Such finds are 'fitted into' their evolutionary history according to the sequence that evolutionists have already worked out for them. For example, if we dig up a number of animals that appear to be closely related, how do we know if they have evolved one into another, or if they are just different varieties of

---

39 For example, the platypus often falls into this category, because there is no living organism that it could have conceivably evolved from or into!

30

the same species? In this case, it is one's *assumptions* that govern the interpretation of the evidence we find.

⊠ Horse Evolution – Horse fossils do not show a clear progression over time as we are told, because many of them are in fact found in the same layer. Sometimes, some of the so-called 'earliest horse' fossils are found right next to the 'modern' horses! What we actually see in the fossil record is no more than variation in the category of horses - and this variation is little different to what we see today! For instance, the largest horse alive today is the Clydesdale, and the smallest is the Fallabella (43cm). If we were to dig these up as fossils in a thousand years' time, would people argue that the Fallabella evolved into the Clydesdale? Interestingly, there are also regulatory genes present in extant (currently living) horses today which are able to switch on and off features such as extra toes. It appears that clever genetic programming has fooled some people into believing that the extra toes were lost through Evolution.[40]

Transition Fossils – The idea of one species turning into another is an audacious claim, and evolutionists are keen to support that claim by providing examples of species which are placed between major families because they share features of both. This would provide credibility to the claim by being a record of one species having turned into another. Two common examples are:

⊠ Whale Evolution - *Pakicetus* – The discoverer (Gingerich) claimed that *"In time and in its morphology, Pakicetus is perfectly intermediate, a missing link between earlier land mammals and later, full-fledged whales."*. What was actually found was a few skull fragments of a wolf-like creature that allegedly had an inner ear like that of a whale. *Ambulocetus* – this has been presented as a walking whale that swims, and was complete with pictures, but all that was found was whale bones with the rear leg bones being located 5 meters above the rest in the rock strata. Since the pelvic girdle was not preserved, there is no evidence of a connection between the hind limbs and the axial skeleton. It is also dated more recently than true whales and is therefore unlikely to be ancestral (by evolutionary dating). *Basilosaurus* – was a 20-meter-long serpent-like sea mammal that was fully aquatic, so can hardly be considered transitional. The small non-functional hind limbs are thought to be necessary for mating – which is also the case for snakes.[41]

⊠ Archaeopteryx - Has all the features indicating that it was truly a bird. Considered by some as a mosaic or chimera (as with the platypus) – which is more in support of Creation than Evolution, because it shares

---

40 Jonathan Sarfati *'The non-evolution of the horse'*, and Peter Hastie *'What's happened to the horse?'* @ creation.com.
41 Jonathan Sarfati *'Refuting Evolution'* 1999 and *'Refuting Evolution 2'* 2002.

features with other organisms not in its 'evolutionary family' (phylogeny).[42]

☒ Dr Colin Patterson, who was the senior paleontologist (fossil expert) at the prestigious British Museum of Natural History, wrote a book on Evolution. When questioned as to why he had not shown any photographs of transitional fossils in his book he replied "*If I knew of any, fossil or living, I would certainly have included them... Yet Gould and the American Museum people are hard to contradict when they say there are no transitional fossils... I will lay it on the line—there is not one such fossil for which one could make a watertight argument.*"[43]

Overall, fossils do not provide evidence for Evolution. Even given the evolutionary time-scale, most fossils appear abruptly in the record without any lineage, and then do not change throughout their entire geological history. The fossil record supports creation of broad groups of animals which have programmed adaptability within genetic boundaries. If the fossil record is to be considered a sequence of events through time, then all it records is non-evolution. Many organisms that have been presented as an evolving series have that appearance because evolutionary dating methods place them into a chronological sequence, rather than presenting these as simply varieties within a kind. However, without the evolutionary time-scale[44] running through the fossil record, the fossils present strong evidence for a global flood - which alone provides the necessary conditions to bury billions of plants and animals all around the world under sedimentary rock.

Ape-Men

So-called 'missing links' between man and apes are used to support the evolutionary theory. These can appear fairly convincing on glossy magazine covers, but what usually remains a secret from the general population, is that actual evidence offers very little support for Evolution.[45] Below are some examples of popularized 'ape-men', and some of the updated evidence pertaining to them.

Invalid Taxons – these are fossils that are no longer considered to be a real species:

☒ Piltdown man – was discovered to be a fraud only after it had been

---

42 See Jonathan Sarfati '*Refuting Evolution*' 1999 and '*Refuting Evolution 2*' 2002.
43 Sunderland, L., *Darwin's Enigma*, Master Books, Arkansas, USA, pp. 101–102, 1998. Patterson's letter was written in 1979. See article '*That quote!—about the missing transitional fossils*' http://creation.com/content/view/5543/.
44 This time scale is addressed in Chapter 5.
45 See www.creation.com Question and Answers, and in particular the peer-reviewed paper by Peter Line: *Fossil evidence for alleged apemen – Part 1: the genus Homo*, and *Fossil evidence for alleged apemen – Part 2: non-Homo hominids* @ creation.com.

used as evidence for Evolution for more than 40 years. It was a human skull combined with a Chimpanzee jaw which had been stained and filed down to look more human.

☒ Nebraska Man – artists depicted this fossil as an ape that walked upright – but the body shape, head, nose, ears, hair, wife, domestic animals and tools were entirely a product of imagination, because the only thing they found was a single tooth – which now turns out to be from a pig!

☒ Homo Habilis – often considered to be an 'invalid taxon' (never existing) because it was made up of more than 300 bones. Nevertheless, it was considered to be a type of ape that stood about a meter high.

Apelike Apes – These are fossils of creatures which distinctly fall into the category of 'Apes'.

☒ Sahelanthropus tchadensis – the most likely conclusion now appears to be that this was a type of gorilla no longer living, distinguished from living apes because of smaller canine teeth.

☒ Orrorin tugenensis – consisted of 13 pieces of bone including broken femurs, several teeth, and bits of the lower jaw. It supposedly walked on 2 legs because of a human-like femur, but the evidence for this is sketchy at best. There was no way of knowing what the head looked like, and the only real claim it has to being a missing link is its supposed age.

☒ Australopithecus ramidus kadabba (Ardipithecus kadabba) – Made cover of *Time*, and was based on 11 fossil scraps from at least 5 individuals in 5 locations. Thought to walk upright because of a foot bone, even though this was found 16km away from the rest, and supposedly 200,000 years younger - according to their own dating methods.

☒ Ardipithecus ramidus – despite being on the cover of *Nature*, this fossil was comprised of only postcranial, dental and cranial scraps which show a "host of characteristics usually associated with modern apes", in the author's own words. Other reviewers say that the teeth and upper arms are what you would expect from a fossil chimpanzee.

☒ Australopithecus afarensis (Lucy), and Australopithecus Africanus – Lucy was a 40% complete skeleton, but did not include the jaw, most of the skull, hand or foot bones. Had the brain size of an ape, skull and body shape of an ape, and was specialized for climbing trees and knuckle walking. Not considered to walk upright and is classified as a type of pygmy chimp on a side branch which does not lead to man.

- ☒ Australopithecus africanus - Evolutionists have suggested that africanus and afarensis should be considered subspecies of a single species. CT (computer tomography) scans show the bony labyrinth of the inner ear showed the semi-circular canal indentations in crania similar to that of extant (currently living) great apes. Africanus had brain size, skull appearance, body shape and size similar to modern apes. Considered a side branch.

- ☒ Australopithecus Robustus / Paranthropus Robustus, Australopithecus Boisei (Nutcracker man) – From the skulls of these creatures it is obvious that there was nothing human-like about them, and evolutionists regard them as not ancestors of humans, but on a side branch that has died out. Analysis of the inner ear shows they did not walk upright.

Humanlike Humans – these are fossils of creatures that distinctly fall into the category of 'humans'.

- ☒ Homo Ergaster (Turkana boy) – classified now under Homo Erectus

- ☒ Homo Erectus (Java man, Peking man, Solo man) – walked upright, enlarged brow ridge, brain size smaller but within today's range. Both morphology and associated archaeological and cultural findings suggest that Erectus was fully human and should be classified under Homo Sapiens.

- ☒ Archaic Homo Sapiens (Homo heidelbergensis) – within the range of people living today, viewed as simply a race of modern humans.

- ☒ Homo Neanderthalensis – same features as modern humans except more robust and stooped over. Now known to be most likely caused by rickets. Neanderthals were fully human, able to speak, were artistic and religious, and mtDNA was similar to modern human variations.

Overall, the fossil record shows two basic groups of 'ape-men':

Apes	Men
These fossils are presumed to be ancestral of humans because their 'geological age' pre-dates humans. Many evolutionists now acknowledge that these finds are more likely to simply be varieties of extinct animals, and are not considered to be ancestral of humans. None can be effectively linked to part of human ancestry.	These human fossils are suspected as being transitional because of the 'geological' age, and various peculiarities (such as a brow ridge). In general however, they show no more variation than what we see in the human population today.

Aside from their assigned geological age, there is nothing to suggest that either of these two groups represent anything significantly different from what we see today. This, then, is an example of different breeds within a basic kind of animal, but there is no evidence supporting the evolving of apes into humans.

"[I]f fossils attributed to erectus were not those of 'apemen', but fully human, then the case for human evolution essentially collapses, as there is an unbridgeable morphological gap between the australopithecine apes and erectus humans, with no missing links in between."[46]

Of Man and Ape

One puzzling question that remains is, if God does exist, then why would he create apes so similar to humans? Surely if humans had no close relatives, it would be far more difficult for people to believe we evolved from apes. Evolutionists will often present the anatomical and genetic proximity between man and apes to support the argument that it would be possible for one to evolve into another. Now consider this from another perspective: imagine that apes were 99.99% physically identical to humans. This would actually prove that we were much more than evolved apes! If we were identical to apes physically, then this would demonstrate that the differences between apes and man were non-physical! The human experience, characterized by morality, love, and reason etc, must therefore be a product of something immaterial and far deeper than can be acquired through physical changes. If God made apes significantly different from humans (or no apes at all), it would provide an insurmountable obstacle for Evolution to cross. On the other hand, if God had made us very physically similar, that would prove there was something immaterial and supernaturally-created about us. What we do in fact see is that apes are both very similar to humans and also have some significant physical differences[47], which are not sufficient to account for the differences in experience. Thus, what we see in nature is the optimum configuration of similarities and differences we would expect if God had *designed creation to look unlike Evolution.*

---

46 Pg 4, Peter Line *Fossil evidence for alleged apemen – Part 1: the genus Homo* @ www.creation.com.

47 For example, one of the major differences between man and ape is the inner ear of humans, which provides a balance sensor enabling us to walk upright. This is a complicated device, and is very difficult to explain by random changes alone.

Biological Design

Many people have puzzled over why so many things look like the result of Evolution if God created the world. Is God trying to trick us, so that we have more faith?[48] The answer is that creation looks exactly the way it ought to look if it was intentionally designed to look like the work of a single, infinitely creative designer. Nature appears to us as not just fit for survival, but also shows characteristics we would expect if someone had designed it to look *unlike Evolution*. The infinite variety of different species shows an infinitely creative source, whilst the incredible unity of all living things shows that there is a single source. Programmed genetic survivability is evidence of sophisticated design and forward thinking. Darwin and his ilk use imperfections to argue that a perfect designer does not exist, but this conclusion ignores the 99.99% perfection we can observe in nature.[49] In addition, everywhere we look we see creation as filled with a mind-blowing level of sophistication. Logically then, we would naturally arrive at the idea of a creator God from observing creation - if not for someone forcing a naturalistic conclusion from a preconceived point of view.

Without being educated to accept biological Evolution, we would naturally arrive at Creation from even a naive glance at the world. We are educated to filter the evidence through an evolutionary way of thinking until everything in the world seems to support the conclusion that has already been assumed. However, as this chapter has shown, that way of thinking is simply misleading. We can challenge macro-evolution and descent with modification and clearly show that it is not the best explanation of the origin of species. In fact, when weighed up, all of its supporting arguments are evidence favouring biblical Creation!

Biblical History

Biological Evolution has gained almost universal popularity, and is now the grid through which almost all scientists think about biology. However, this has nothing to do with the weight of its supporting arguments, but that it offers a secular scientist the opportunity to examine the world through a Naturalistic/Atheistic Worldview. This is attractive because it opens up the sciences to anyone – not just those who believe in God. However, Evolution is merely a paradigm with which to interpret the facts through a philosophical framework of Naturalism – which has already

---

48 Chapter one of 'The Biotic Message' deals with these issues in depth. Walter James Remine here goes into depth as to why creation looks as it does, and argues convincingly that creation was designed as a message, which by its very nature, is anti-evolutionary.

49 Even though according to the Bible, there is a 'curse' on creation which means that God allows nature to run its course without intervention. Genetic mistakes will accumulate over time, and thus a biblical creationist does not argue that nature is now perfect (only originally perfect).

been shown to be woefully inadequate to successfully explain the facts.

An alternative way of thinking, which has already been alluded to here, is to see biology in terms of biblical history. The Bible claims to record history all the way back to creation, and as already shown in this chapter is very effective at explaining many major aspects of biology – certainly more than the evolutionary alternative. It might therefore be useful to present a short summary of what the history of biology would look like in terms of biblical history:

(1) There was a creation event about 4000BC, when all life was designed and brought into existence supernaturally from an infinitely intelligent single designer. Life was therefore designed for survival and to look like the work of a brilliant and beautifully creative mind (in the same way that an artist expresses themselves in a painting). Life was also designed to be fit for survival through the course of time in a variety of different circumstances. All animals were originally created vegetarian.

(2) There was a global flood which killed all animal life, except for those on Noah's Ark (and those that did not breath through nostrils, such as insects and dolphins), the biblical specifications of which indicate a huge structure with more than enough capacity to house two of every basic kind of animal on the planet (possibly about 16,000 individual animals). God brought (clearly a miraculous event) two of every basic kind of animal onto the ark. Those outside the ark were entrenched in the water and sediments which were stirred up and subsequently caused widespread fossilization of those animals in sedimentary rock, together with formation of coal and oil seams.

(3) After the flood, animals and people (after Babel) began to spread across the Earth and find suitable habitats to live in.[50] The Bible suggests that at this time animals and people became carnivorous, and animals became afraid of man. Programmed genetic variability in plants and animals allows for adaptation through time, which gives them the ability to be suited to different environments, and over time they migrated and filled every niche on the Earth with life. Accumulated genetic mutations over time damaged the DNA leaving an accumulation of defects in the population etc. Extinction of various kinds over time also occurred due to changes in predation (by man and beast) and environment.[51]

---

50 It is thought by many that there was an ice-age at this time when there was unprecedented precipitation at higher latitudes.
51 Many believe that the pre-flood environment was a richer more fruitful place to live. It is thought that after the flood the lifespans of all animals reduced as the climate became harsher. Nowadays, there are places where it is difficult to find water (deserts etc.) and/or sufficient vegetation for species to thrive as they once may have.

Chapter 3 - Abiogenesis

Abiogenesis is the concept that life can come from non-life, and is the first obvious step in the process of Biological Evolution. The general concept is that non-living chemicals turned into some form of self-replicating organism through exceptional circumstances.

The Law of Biogenesis

People once believed that life was relatively simple, and therefore it could arise from virtually anywhere given the right circumstances. For instance, people observed maggots coming out of rotten meat, and assumed that whenever rotten meat was around, life would just appear in this way. This view is no longer accepted by Science, because it is known that in each case of 'spontaneous generation', the life was already there in some form or another. Evidently, flies were getting into where the rotten meat was kept and laying their eggs. Hence the law of Biogenesis was formulated, and this states that life only comes from life. Additionally, the offspring are always of the same kind as the parent. One of its foremost contributors was Louis Pasteur. He was a Christian who believed that no new living things were created/evolved after the creation event (Genesis), and this belief led him to eventually realise that food spoilage and infectious sickness could be prevented, because whatever was causing this was already living and could therefore be killed. In the medical profession, the Law of Biogenesis was a breakthrough, because it showed that some diseases (infections) were a result of an outside influence, and therefore were preventable (the washing of hands) and treatable (antibiotics). Hence Pasteur's discoveries (such as the pasteurization of milk and wine, and the treatment of rabies by antibodies) are based on the concept that only living things can give rise to living things, and form the basis of microbiology, immunology and the food preservation industry today. Food preservation works by merely maintaining a sterile (bacteria-free) environment for a given product, either by canning or the use of preservatives etc. If we took this technology to an extreme, we know that if we were somehow able to sterilize the entire Universe, never again would there be any kind of life.

The reason why Biochemists, Doctors and Food Scientists etc can work on these sort of assumptions is that life is an exceedingly complex and delicately balanced array of chemicals and micro-molecular machines. Therefore, it is very reasonable to suggest that ordinary chemicals cannot self-organize themselves into the kind of complexity needed for life. No counter examples have ever been found to this principle, except for the claim by evolutionists that this is how life originally began.

Without an external life source (i.e. God), Naturalists must believe in the *spontaneous generation* of life from non-living chemicals (in violation of the scientific Law of Biogenesis). To summarize the issue: there is simply no evidence for such a view, except for that fact that life exists today. Obviously no one was there to observe the event, so evolutionary origins science relies purely on speculation, which so far has produced no credible explanations. Evolutionists believe in Abiogenesis because of backwards reasoning: working from the basis that because (1) life exists and (2) God does not exist[52], then spontaneous generation must have occurred at least once in the past (even though it does not happen today).

Darwin himself speculated on the origin of life: "*But if (and oh what a big if) we could conceive in some warm little pond, with all sorts of ammonia and phosphoric salts, light, heat, electricity, etc., present, that a protein compound was chemically formed...*"[53]

Spontaneous generation of life from non-living chemicals is obviously the first critical step in the evolutionary tree, and without a plausible scientific scenario, the entire naturalistic framework rapidly collapses in on itself. Whilst a number of scenarios have been suggested, scientists are still a very long way away from finding *any* naturalistic scenario that is *remotely plausible*. For this reason, the origin of life is still widely regarded as a mystery. Even the famous scientist Sir Francis Crick (co-discoverer of the structure of DNA) said that "*The origin of life appears to be almost a miracle*".[54]

(1) An Electric Spark - Scientists propose that the early atmosphere contained 'reducing gases' such as carbon dioxide, water vapour, nitrogen and possibly methane and ammonia. It is argued that the basic organic molecules necessary for life could have formed in this gas with the input of electrical energy or ultraviolet light. Experimental support for this idea arrived in the 1950s with a very famous experiment by Stanley Miller. He designed apparatus to send an electrical spark through a mixture of gases (ammonia, hydrogen, methane and water vapour) designed to emulate what they thought Earth's early atmosphere might have been. The result was that several amino acids were formed, and in later similar experiments, other scientists have synthesized different mixtures of amino acids, sugars and nucleic acids (needed for DNA) as well.

☒ The production of a few of the building blocks of life can hardly be called evidence for a scenario of Abiogenesis, because that would be akin to saying since some rocks are naturally square enough to stack,

---

52 There remains no effective argument supporting this. Evolution merely assumes no God.
53 Francis Darwin, *The Life and Letters of Charles Darwin* (New York: D. Appleton, 1887), pg 202.
54 Francis Crick, *Life Itself* (New York: Simon and Schuster, 1981).

skyscrapers can construct themselves out of piles of rocks given enough time. The reality is that life is far more complex than this, and this cannot be explained by acquisition of merely a handful of the vital chemicals.

☒ Miller-type experiments do not accurately reflect possible conditions. For instance, the experimental apparatus is designed to remove and isolate the valuable amino acids from the reaction chamber where they would otherwise be destroyed by the electrical discharge that made them (electricity is harmful to living things). They are also unrealistically isolated from other harmful chemicals (and concentrated). Such a carefully contrived experimental set-up is not an accurate simulation of reality.

☒ Many scientists now acknowledge that Earth's early atmosphere could not have contained methane, ammonia or hydrogen – and these gases are necessary for the 'Miller-type' scenario to work.[55] Even granting an old-Earth time-scale (where the atmosphere was not necessarily the same as today), estimates of the early atmosphere say it likely contained water, carbon dioxide and nitrogen, from which no such Miller-type spark experiment would be successful in producing organic compounds.

☒ The early atmosphere likely contained oxygen (still assuming an old-Earth time-scale) which would have destroyed organic compounds (without a cell wall). Even if there was no oxygen, this would mean there would have been no ozone, and the chances of destruction by UV light increase. It appears, then, that life on Earth cannot have begun spontaneously either under a reducing atmosphere (that contained methane, ammonia or hydrogen), or an oxidizing one (the opposite).

(2) Some warm little pond – In agreement with Darwin's own words, there is widespread acceptance of the idea that life came about from a 'primordial soup', which is conceptualized as a random array of chemicals that over time happened to coalesce into the right mixture to form the first living cell.

☒ Aqueous environments are not conducive to polymerization:[56] water is released (a condensation reaction) in the joining of amino acids (a peptide bond), but excess water drives the reaction in reverse and breaks up these polymers as per:

$$\{aa\}\text{-}NH_2 + HOOC\text{-}\{aa\} \leftrightarrow \{aa\}\text{-}NH\text{-}OC\text{-}\{aa\} + H_2O$$

☒ For this reason, DNA too cannot last long outside the protective

---

55 See Jonathan Sarfati *Loopholes in the Evolutionary Theory of the Origin of Life: Summary* @ creation.com.
56 See Jonathan Sarfati (1998) *Origin of Life: The Polymerization Problem*, @ creation.com.

chemistry within the cell. Polymerization also requires the right mixtures of 'bifunctional' monomers (amino acids which can link with two others) with unifunctional ones (which start/stop the polymer chain). All prebiotic simulation experiments produce 5 times more unifunctional units, thereby effectively preventing polymerization.[57]

☒ Even if amino acids were present in a sort of primordial soup, it is highly likely that they would have been destroyed by reacting with other chemicals present.[58]

☒ Life as we know it requires homochiral amino acids.[59] Proteins use only 'left handed' molecules, while DNA & RNA use only 'right handed' sugars. The chirality (specific arrangement in space) of these compounds comes from being manufactured by homochiral enzymes – which obviously were not present before 'left handed' amino acids (because they are made of 'left handed' amino acids)! Normal chemical processes produce mixtures of left and right handed molecules (a racemic mixture).

☒ The building blocks of life are inherently unstable.[60] Many of the key chemicals of life would destroy each other (e.g. reactions between sugars and amino acids) if not for the protective mechanisms within living cells. Ribose and cytosine are particularly hard to form and are very unstable.

☒ The proteins required to synthesize histidine themselves contain histidine!

(3) Submarine Hydrothermal Vents – So called 'black smokers' are hydrothermal vents on the oceans floors where warm water full of dissolved minerals is ejected. Some scientists propose that these sites could possibly provide the right conditions necessary for the beginnings of life. The essential elements for life (C, H, O, & N) are present in the seawater along with other key elements such as zinc, copper and iron. For this reason hydrothermal vents support a number of organisms, some of which were previously unknown. It is believed by some that life began with the polymerization of the kind of amino acids which have been detected in black smokers.

☒ A hot aqueous environment, however, is not chemically conducive to polymerization. It destroys some of the more delicate amino acids (such as serine and threonine), and it racemizes the amino acids into a biologically unfriendly mixture of right and left-handed configurations.

---

57 Jonathan Sarfati *In Six Days* pg 70.
58 *The Case for Faith*, interview with Stephen Meyer pg 228.
59 See Jonathan Sarfati (1998) *Origin of Life: The Chirality Problem*, @ creation.com
60 See Jonathan Sarfati (1999) *Origin of Life: Instability of Building Blocks*, @ creation.com

☒ High temperature increases the rates of a number of undesirable chemical reactions which would destroy biologically important molecules.

☒ It is not possible to form chemical bonds between amino acids and nucleotides in any aqueous environment (e.g. ponds, lakes and oceans), because water rapidly destroys the high energy reagents required to form these bonds.[61]

☒ Stanley Miller himself said that polymers are *"too unstable to exist in a hot prebiotic environment."*[62]

(4) Proteinoids – Some scientists seeking to overcome the difficulties in an aqueous environment have experimented with heating and dehydration. Fox[63] has discovered that when a mixture of pure, dry amino acids is heated to 175°C for a limited time, microscopic globules of amino acid polymers are formed. These have been dubbed 'proteinoid microspheres'. Under the right conditions these proteinoids can aggregate into microspheres resembling a cell-like membrane. Proteinoids are also capable of multiplying by division. These are formed when a concentrated mixture of amino acids is dehydrated and heated, and as a result polymers consisting of more than 200 linked amino acids are formed. The theory is that these microspheres are a sort of 'protocell', representing a kind of precursor to the first life on Earth.

☒ The chemical conditions needed to find a concentrated supply of pure amino acids (heated to 175°C for 6 hours then cooled immediately in water) is a very unrealistic scenario anywhere on Earth. Amino acids produced from simple precursors (Miller spark experiments for instance) would also have a number of other organic compounds associated with them which, when heated together, would certainly destroy the amino acids.

☒ Fox's scenario requires specific proportions of amino acids, which does not represent any naturally occurring conditions. No proteinoids are produced, for instance, when random proportions of amino acids are heated.

☒ Not only does heating tend to destroy some amino acids, but it also racemizes the mixture, resulting in a 50/50 mixture of right and left handed amino acids. Substitution of the D (right) amino acids in

---

61 Duane Gish *Origin of Life: The Fox Thermal Model of the Origin of Life*, pg 8. @ www.icr.org/articles/79
62 Miller, S.L. And Lazcano, A., *The Origin of Life – Did It Occur at High Temperatures?* J. Mol. Evol. 41:689-692, (1995).
63 S. W. Fox, *"The Protein Theory and the Origin of Life,"* American Biology Teacher, Vol. 36, pg 161-172 (1974).

biological systems can completely destroy the function of a given protein, and this shows that proteins *must* be of the same chirality – but there are no natural processes known to generate them.

☒ The 'cell wall' structures are far from those found in biology, and do not have the biochemical machinery to have a chemically controlled environment within them. The apparent replication of these microspheres is again far from that seen in biology, because we are only seeing a *division* and no duplication of the internal components – which requires far more complex cellular machinery and timing.

☒ *"An outline of Fox's theory can be found in practically every modern high school and college text on biology, evolution and related subjects... And yet if any thing in science is certain, it can be said that however life arose on this planet, it did not arise according to the scheme suggested by Fox."*[64]

(5) Biological Predestination – It has been suggested that information-rich biological molecules have certain sequences into which they will naturally self-order, because they are a result of chemical affinities. A famous textbook *Biological Predestination*, popularized this theory.

☒ Dean Kenyon, the author of the above textbook, has since repudiated his own theory, saying *"we have not the slightest chance of a chemical evolutionary origin for even the simplest of cells"*.[65] He even said in the forward of the book *The Mystery of Life's Origins* that *"The authors believe, and I now concur, that there is a fundamental flaw in all current theories of the chemical origins of life."*[66] As an expert in the subject of self-ordering molecules, his view is very damning to the theory.

☒ Besides the fact that no significant affinities have been found in amino acids (at least not ones that correlate to actual known proteins)[67], such affinities would cause the amino acids to sequence in the same ways over and over, and therefore they would not be able to form into all the different kinds of proteins necessary for life.

☒ Since the structure of DNA involves amino acids attached to a sugar-phosphate backbone, there is no bonding between the amino acids themselves. Therefore, this isolation eliminates any affinity they would have for a particular order. Even if this was to happen, it would defeat

---

64 Duane Gish offers a devastating critique of Fox's proteinoids in: *Origin of Life: The Fox Thermal Model of the Origin of Life*, @ www.icr.org/articles/79
65 Interview in *Unlocking the Mystery of Life*.
66 Charles B. Thaxton, Walter L. Bradley, and Roger L. Olsen, *The Mystery of Life's Origins* (Dallas, Tex,: Lewis and Stanley, 1984).
67 Interview with Stephen Meyer in *The Case for a Creator*, pg 233.

the purpose of having DNA as an information carrier, and hence it would not be able to account for the biology that we observe. If certain sequences were a result of pre-determined chemistry, which would be like buying a blank CD on which you could record only one particular song!

(6) Clays – It is thought that the combination of biological molecules joining together into information rich bio-polymers could occur more readily on the surfaces of clays, which would provide an environment that is more conducive to polymerization and perhaps act as a sort of template for sequencing.

☒ The order that we can observe in certain chemicals is a result of the properties of those chemicals. Therefore they cannot function as a template for the order we see in DNA, because this order is not a function of chemical properties, but is a coded language for functional proteins etc. Highly ordered substances like crystals, if used as a template, could only ever produce repeating patterns, and thus, is not a possible candidate for the kinds of biological molecules that we observe.

(7) RNA – More recently, RNA has been suggested as a primary polymer of life, because it has been shown to be able to self-replicate without the aid of complicated protein enzymes, and it also has the ability to catalyse certain biochemical reactions.

☒ However, the synthesis of RNA would be a lot more complicated than simple amino acids polymers, because RNA requires the synthesis and polymerisation of three molecules: an amino acid, a sugar and a phosphate group. For example, the difficulty in the formation of the amino acids is compounded by the formation of sugars. The sugars required for the DNA/RNA molecule need to be exclusively 'right-handed', and pre-biotic reactions such as the formose reaction produce a mixture of left and right (L/D). Additionally, the same process known to form sugars is also known to then go on to destroy them.

☒ The formation of chemical bonds between the sugars, phosphoric acids, and amino acid base pairs (purines and pyrimidines) requires energy in the form of ATP – which is itself generated by complex chemical pathways. The same energy is released when these bonds are broken, and the natural tendency is for these molecules to fall apart.

☒ *"We cannot say that the pre-biotic synthesis of nucleotides is impossible. We know only that if it happened it happened by some process which none of our chemists has been clever enough to reproduce."* [68] One of the leading researchers in the 'RNA world'

---

68 Dyson, *Origins of Life*, Cambridge University Press (1985), pg 24.

hypothesis said: *"The most reasonable assumption is that life did not start with RNA ... The transition to an RNA world, like the origin of life in general, is fraught with uncertainty and is plagued by a lack of experimental data."*[69]

(8) Panspermia – Is a concept promoted by famous scientist Francis Crick (co-discoverer of the DNA double helix structure). The suggestion here is that because the chances of life forming here by chance are so remote, perhaps it is more reasonable to conclude that more favourable conditions were fulfilled elsewhere in the Universe on another planet, and then somehow that life got transported here where it seeded and evolved. Or perhaps, that another civilization, possibly facing extinction, deliberately seeded this planet with primitive life. The evidence for such an event would be the sudden appearance (in the fossil record) of fully formed complex life – which is exactly what we do find!

☒ In reality, Panspermia is the most unreasonable of explanations. It is pure fantasy, and lacks evidence of any kind. Whilst ignoring the tremendous obstacles in getting life from one planet to another, basically what we see here is an admission that naturalistic science simply cannot propose a realistic set of conditions that had to be met on planet Earth. That being the case, if the chances of life evolving here are so remote, then it would seem even more remote to suggest that they evolved elsewhere, given that the chances of life evolving are compounded by the chances of there being another life-friendly planet like our own out there somewhere.

☒ It is also very unlikely that any living cell could survive the immense heat generated on impact with Earth.

☒ Of course, suggesting an alternative location doesn't actually get around the problem - it just shifts it elsewhere. There is still no explanation for how the chemistry of Abiogenesis would work. For example, the complex inner workings of the cell are still going to be complex both here and on Mars, and the DNA code is going to be just as unlikely to come about by chance anywhere in the Universe.

☒ *"When a scientist of Crick's calibre feels he has to invoke undetectable spacemen, it is time to consider whether the field of prebiological evolution has come to a dead end."*[70]

Above are just a few notes showing that each scenario envisioned by evolutionists (and stated as fact in textbooks) can be found to contain critical errors. Experts in those fields are aware of many more, and in every area of investigation scientists are scrambling to discover a plausible

---

69 Joyce, G. F., *RNA Evolution and the Origins of Life*. Nature, 338:217-224 (1989).
70 Phillip Johnson *Darwin on Trial*, pg 111.

model explaining the mystery of life.

"The mood at the 1999 international conference on origin of life was described as grim – full of frustration, pessimism, and desperation. Nobody pretends that any alternative provides a reasonable path of how life went unguided from simple chemicals to proteins to basic life forms."[71] One journalist sums it up saying: *"Science doesn't have the slightest idea how life began."*[72]

Life by Design

The Design Argument[73] has been around for a long time, but more recently it is gaining momentum from recent scientific evidence. It can be articulated as follows: *Where there is evidence for design, there must be a designer.*[74] As scientific investigation discovers more of the intricacies of life and the Universe, what we are discovering is there are more and more elaborate and incontrovertible evidences for design everywhere we look. In particular, the relatively new science of Biochemistry is revealing that what we call 'life' is in fact a staggeringly complex array of chemicals and macromolecular machines that form cleverly operating chemical systems.

Through progressive discoveries of the inner working of the cell (the basic unit of all life) we now know *why* the law of Biogenesis makes sense. Every living cell is like a microscopic city of mind-boggling complexity: it is filled with a bristling array of thousands of chemicals with tiny pumps operating to control their concentrations, thousands and thousands of tiny machines each performing a specific role in a elaborate biochemical chain of processes, an automated production line made of machines assembling other machines, cell components and chemicals, and a centralized computer system that holds the information to build, control, repair, defend and replicate the entire system. The cell is divided into sections where different organelles perform their various tasks: the mitochondria; to produce energy, the endoplasmic reticulum; to manufacture proteins, the Golgi apparatus; to hold proteins, the lysosome; a waste disposal system, secretory vesicles; a storage facility, the peroxisome; to mobilize fats, and the cell nucleus; to house the centralized information storage and retrieval system which controls the rest of the cell. And this cell in all its complexity performs only its specified role in the overall integrated chemical system which makes up the organism.

The human body, for example, is made up of 75 trillion cells, all assembled

---

71   Interview with Walter Bradley in *The Case for Faith*, pg 150.
72   Gregg Easterbrook, *The New Convergence*. Quoted in *The Case for a Creator*, pg 41.
73   Also known as The Teleological Argument for the Existence of God.
74   The real question being how to identify design. Usually this is obvious, but anything that has a low probability of being formed by chance, coupled with showing for example: high technology, complexity, interdependent parts, precise or delicate construction. Such things show forethought in their construction.

into an ordered system, where each one performs its specialized function in the anatomical position they are assigned. The technology required for even a 'simple' organism to be self-sustaining and self-replicating is more complex and precise than anything that we humans have ever built. The fact that such a technological marvel can be compacted into a microscopic unit is testimony to its engineering genius. Furthermore, that all this life can be expressed in terms of a single set of coded instructions is a feat of digital programming that is beyond our ability to even comprehend. It is extremely doubtful we will ever be able to come up with a code that, when read, turns itself into an autonomous system capable of controlling, repairing and self-replicating itself!

Thus, because the complexity of the basic unit of life demonstrates a high degree of technology (with a low chance of self-assembly), it follows that there must be an intelligent designer for that life. Naturalistic explanations for the origin of life fail to account for two major observations: (1) The complexity of the biochemistry, and (2) The complexity of the DNA code. It is helpful, for this discussion, to liken life to a modern computer. For a computer to be running as we are used to, it requires both functioning hardware and software. The hardware is the physical structure of the computer which allows for programs to be run, and is comparable to the physical aspects of the living cell (the vast array of chemicals and macromolecular machines). The software of a computer is the (metaphysical) programming that the computer runs, and can be compared to the coded message found in the DNA of every living cell. Thus, both a computer and a living cell need both the 'hardware' and the 'software' present on start-up for operation. The difficulty for Naturalism is that we are now finding that the cells of the simplest forms of life require both super-computer hardware and super-advanced programs to be present for life to exist, and that this technology simply cannot be accounted for by anything other than a super-intelligence. For example, consider that the human brain is understood to be the most complex structure in the Universe, capable of forming a whopping 100 trillion different synaptic connections through some 400,000 km of nerve fibres. This would be far more complex than the combined connections of the entire Internet, but yet this system is somehow manufactured from a simple code in the DNA. Both the programming and physical processes of the cell must comprise a very advanced manufacturing system to be capable of producing such a technological marvel from simple coded instructions.

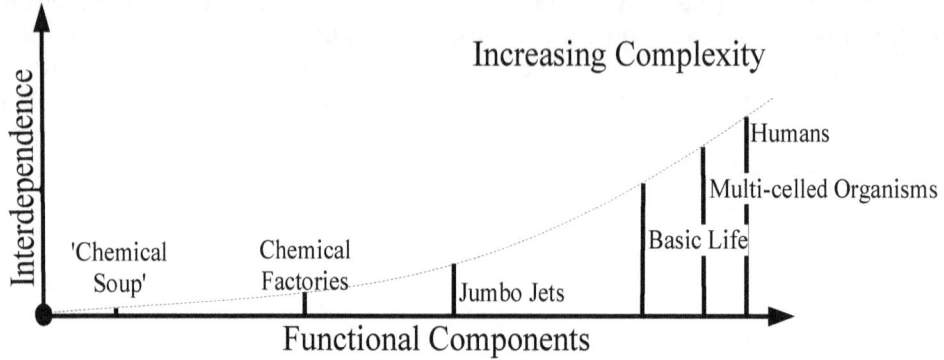

"The simplest organism capable of independent life, the prokaryote bacteria cell, is a masterpiece of miniaturized complexity which makes a spaceship seem rather low-tech."[75]

-The Hardware Problem-

What we call 'life' is basically a fragile arrangement of chemicals which function together to sustain their own existence, for example by producing their own energy, repairing and reproducing themselves as an integrated system. This basically requires the presence of three major classes of key molecules confined within a boundary wall:

(1) DNA (Deoxyribose Nucleic Acid) – This is the central database of information that controls everything else. In a single-celled bacteria such as *E. coli* there are about 4,600,000 nucleotide base pairs, and in humans there are just over 3 billion base pairs. That is enough to fill more than 1000 five-hundred page books with coded information. The DNA is a very long polymer (about 6 feet in every human cell!), and it is counterbalanced by an equal and opposite strand running in the opposite direction in a combined structure known as the double helix (this provides a backup copy of the information). The DNA is found tightly wound up and compacted into manageable structures we call chromosomes. The strands of the DNA are a repeating trimer unit consisting of a phosphate, a sugar called *ribose*, and one of 4 amino acids (*adenine, guanine, cytosine,* and *thymine* or AGCT).

(2) RNA – When DNA is unwound and copied into a single-stranded fragment, we call it RNA. This is a sort of information carrier molecule that the cell uses to transfer information from the central database of the DNA into the rest of the cell to control its various subfunctions. For example, the RNA is used as a base template to code for all of the various proteins that need to be manufactured in the cell throughout the course of the cell's life. These are made by translating the code in

75 Phillip Johnson *Darwin on Trial* pg 105

the RNA into the corresponding amino acid sequence. Each set of three nucleotide bases in the RNA code for one amino acid, and so a strip of RNA of, say, 600 units long, would be needed to code for an average protein containing 200 amino acids.

(3) Proteins – Most of the components of the rest of the cell and the overall organism are made up of different sorts of proteins. There are virtually an unlimited number of different machines and micro-structures that can be built out of protein, because there are so many different shapes possible, depending on the specific sequence they are ordered into. Amino acid polymers (the basis of proteins) are incredibly versatile because their sequence causes them to fold into a very precise shape. Often, proteins are folded in such a way as to host a coordination complex which functions as an 'active site'. These proteins, called enzymes, are responsible for all the chemistry that occurs within the cell because they house specific metal salts in order to each catalyse a particular chemical reaction.

Protein Assembly

Proteins operate like tiny machines, and together perform all the functions of the cell. All the chemical processes which produce or break down the thousands of chemicals in the cell (for example, energy production) are done by different types of proteins and enzymes. The manufacture of these proteins is achieved in the Protein Synthesis Apparatus (PSA), which is a set of at least 75 different macromolecular machines[76]. The first step in the process involves the unwinding and transcription of the DNA to produce the messenger RNA fragment. This fragment is then transported out of the cell nucleus and into a two-part protein assembly factory called a ribosome. Here, amino acids which have been transported in from other parts of the cell are matched up with the corresponding nucleotide bases to add to the growing polymer. When all the RNA has been decoded and the amino acid chain is completed, it is then transported into a barrel-shaped machine that helps it fold into the correct 3-dimensional conformation. It is then shepherded by another molecule into the location where it is needed. Each of the proteins involved in this process are in themselves precisely engineered machines put together *in the same factory*.

Thus, it takes many precisely engineered proteins to produce proteins. That is, the PSA is made up of proteins manufactured by the PSA. The DNA blueprint for life cannot be read unless there are all 75 of these proteins present - which in themselves are a product of DNA having been read. Therefore, both complex protein machines and the correct DNA sequence would need to be present in a theoretic first cell. All of the other

---

76 John Marcus, *In Six Days*, pg 161

chemicals and energy molecules (ATP) necessary for this process and to keep the cell alive are also synthesized by proteins created by the PSA. Obviously, the 75 machines that make up the PSA could not function unless all were held together in a controlled environment together with the DNA. In order to achieve this in a living cell, there is a membrane or cell wall. The conditions inside the cell are controlled by a number of different kinds of proteins and ion pumps, which allow specific chemicals in or out of the cell. Again, because these are made up of protein, they could not exist without the PSA, which itself could not exist without the conditions that the inter-membrane proteins create. In terms of a theoretical first living cell, all these systems containing functioning proteins, RNA and the correct sequence of DNA would need to be in place before the cell could self-sustain and self-replicate.

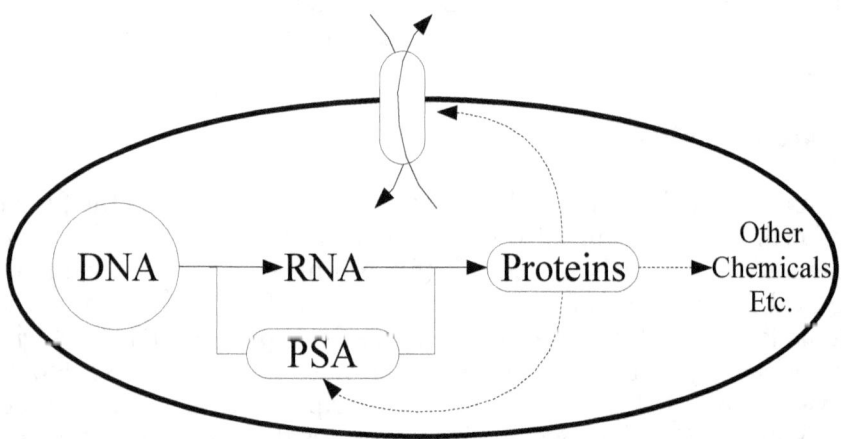

Life therefore requires a large minimum number of complex working parts, and therefore cannot be built up one component at a time. Only a parent cell which has divided can have all the necessary start-up requirements for another living cell. Thus, Biochemistry gives a rational basis for the law of Biogenesis, and explains the observation that only life ever produces life.

"A cell can only come from a functioning cell and cannot be built up piecemeal, because all the major organelles must have been created and assembled instantaneously for the cell to exist."[77]

A theoretical first living cell would have to be vastly complex before it could self-replicate. Therefore, the incredibly unlikely convergence of chemicals and protein machines would need to be achieved simultaneously in order for the possibility of a successive generation. Life

77 Jerry Bergman, *In Six Days* - in reference to Dean Overman *A Case Against Accident and Self-Organization*, New York: Rowman & Littlefield Pub, 1997.

cannot have begun any simpler than a very complex cell. Darwin himself said that: *"If it could be demonstrated that any complex organ existed which could not possibly have formed by numerous, successive, slight modifications, my theory would absolutely break down."*[78] When this statement was made in 1859, the scientific community had no concept of the actual complexity of life. However, it now turns out that we are hard pressed to find any such "complex organ" which *can* in theory be a product of stepwise progressions. Two observations support this conclusion: (1) That most biological systems that we can consider as examples possess such a high technology that they must be almost perfect to have any function whatsoever, and (2) That these biological systems are also dependent on the perfect function of numerous other systems. It seems as though almost all biological systems therefore require a large number of minimum parts in order to have any function at all. This also gives rationale to why there are so many different ways the human body can get sick, because it only takes the failure of one small component for the overall system to collapse. It is the same with a car, which will break down and become useless if only one of the many functioning parts fails.

Biochemical pathways are a good biological example: living organisms are essentially (physically) made of complex chemical mixtures, protein machines and structures. Each chemical the body needs (or needs to break down) is manufactured in a chain of chemical reactions, where the overall chemical transformation occurs through many individual and successive steps. Living things utilize catalytic technology in precisely engineered proteins called enzymes. These molecules typically are a specific sequence of amino acids, which fold into a particular shape to form a structure to house a particular metal. The 'coordination complex' formed by the protein (which functions as a ligand) and the metal together make an 'active site' which reacts with a particular chemical converting it into another chemical. These enzymes are in themselves very complex machines, representing a far more advanced and environmentally friendly technology than anything created by man. They react only with the correct chemical, they produce no by-products, they use little or no energy, they are water soluble, non-toxic, can produce very complex compounds with an incredible precision (with the correct chirality for example), and can be produced rapidly in large numbers with their concentration accurately controlled. The second thing to note is that most biochemical pathways involve a chain of enzymes, with each one performing its specific function in the right sequence of events. Each enzyme is therefore useless until all the other enzymes are in place.

One such example is the Citric Acid Cycle (pictured below), which is an important biochemical cycle that uses *pyruvate* as the feedstock material to produce energy in the form of ATP through the electron transport

---

78 Charles Darwin, *The Origin of Species* (New York: New York University Press, sixth edition, 1998), pg 154.

system. The overall system is a critical biochemical system for life, and features a chain of complex enzymes which function in a chemical cycle.

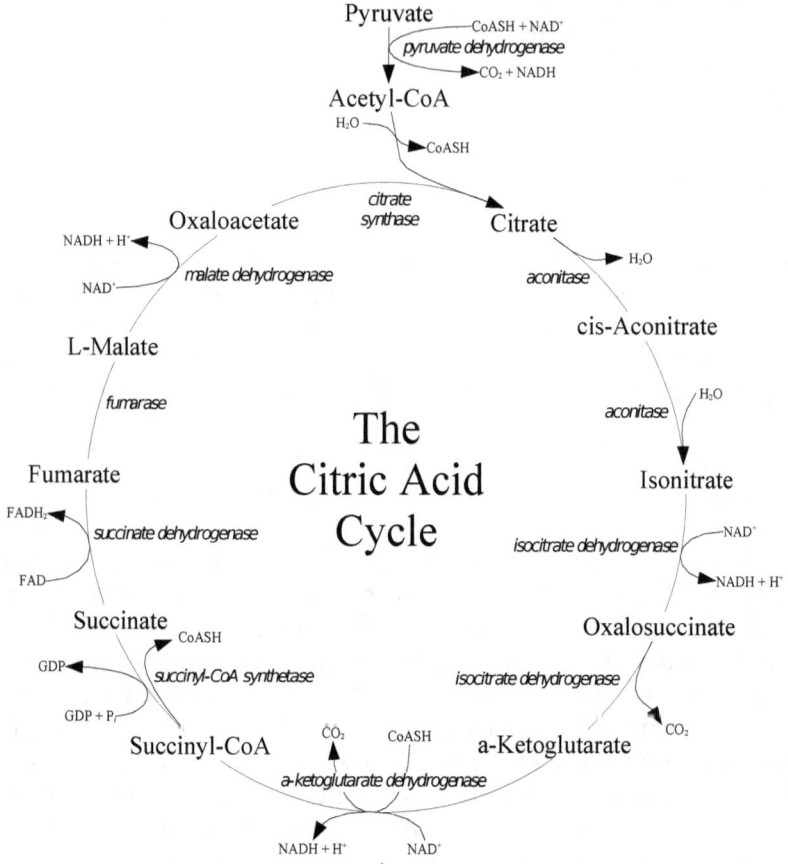

This system is of course just one system that itself is dependent on a number of other chemical systems, such as the biochemical pathway to produce Acetyl CoA and the protein assembly to manufacture the enzymes in this cycle. Each one of these enzymes in such biochemical pathways is in itself a complex machine, and it only takes one malfunctioning enzyme to shut down the entire (interdependent) system, and associated systems. The overproduction of a particular chemical can quickly poison the overall system. Antibiotics, for example, function by shutting down a particular chemical pathway in the bug. The famous British evolutionist J.B.S. Haldane said that Evolution could never produce "*various mechanisms, such as the wheel and magnet, which would be useless till fairly perfect.*"[79] Modern Science has discovered many such structures in nature[80], and biochemical pathways take the precision required for a perfect 'wheel' to yet another remarkable level.

79 Dewar, D., Davies, L.M. and Haldane, J.B.S., (1949). Is *Evolution a Myth?* A Debate between D. Dewar and L.M. Davies vs. J.B.S. Haldane, Watts & Co. Ltd / Paternoster Press, London, pg 90.
80 Jonathan Sarfati, Design in living organisms (motors), (1998) @ www.creation.com

The eye is another obvious example: The eye is made up of a number of functioning parts, each of which must be working correctly. There is the cornea, the iris, the lens, the fluid within the eye, the retina, the optic nerve and the visual cortex within the brain. If any one of these systems fails, then the overall system loses function. Now, each of these components within themselves employ a very high degree of technology to achieve their function, and require a large number of correctly operating macromolecular machines, which are themselves very complex and interdependent. The biochemistry, for example, that converts light into an electrical signal in the retina's 100,000,000 light sensitive cells is extraordinary, and an example of a technology far more advanced than optical systems we have created. Each of these precisely engineered macromolecular components must be functioning correctly for the part to have a function, and yet each functioning component is interdependent on the other components of the system. Thus, this is an example of a system that cannot be reduced in any way without total loss of function. The system is 'irreducibly complex'[81] and relies on individual components which in themselves are irreducibly complex. There is also a tiny muscle on the base of the eye which causes the eye to vibrate at between 30-70 Hz. This continual but slight movement of the eye causes the incoming light to be continually striking different optical receptors on the retina, and essentially means that the image updates 30-70 times a second. Without this tiny muscle working effectively, everything except rapidly moving objects would be invisible to the eye, because the optical receptors would so rapidly become saturated by the image that it would disappear. Thus, the eye gives you an idea of how complex and delicately balanced such biological systems can be.

Such systems (almost every biological system) and biochemical pathways cannot be arrived at by the addition of one component to another, because the function of each component is reliant on the function of all the others. Therefore, no component can be constructed without knowledge of the function of all other components, and in the same way as with a car, the overall system needs to be planned conceptually in advance *before* individual components are made. The implications of this are that all biological systems that demonstrate irreducibly complex and/or interdependent systems must have involved prior conceptual design *before* construction. Irreducibly complex systems are the norm in biology, and there would be thousands of examples one could give to demonstrate the complexity of biological systems.

Irreducibly complex systems are essential within the biochemistry of the cell and in complex components such as the eye, but there is also irreducible complexity and interdependence between components in the body.

---

81 Michael Behe promoted this concept in the book *Darwin's Black Box, The Biochemical Challenge to Evolution*, New York: Free Press (1991).

Flight is a good example.[82] Birds are supposed to have evolved from reptiles, according to evolutionists, but this would require the unrealistic *simultaneous* evolution of a number of interdependent components, and until all of the components were fully functional, they would more likely pose a disadvantage to the species. Flight is made possible firstly from a wing of the right shape, size and aerodynamic function. Alongside this complex piece of engineering, however, there would also need to be the simultaneous development of the rest of the system essential for flight. This would include a change in the corresponding nervous system and blood vessels to support the wing, a change from cold blooded to warm blooded circulation, the development of the avian lung to support the higher oxygen demands of flight, more powerful muscles under the wing and a central point of connection (the keel), a new hollow bone structure with added bones for the wing, a forward facing elbow joint and the versatile swivel joint needed for the flapping motion, also the brain function necessary to control flight, and a feathered tail. The feathers themselves are virtually useless without the interlocking barbules which maintain their structure, and the preening gland for lubrication. Obviously the development of such a system would need forward thinking - a feat impossible for random chances. Note also that the enormous number of mutations required to achieve this would need to occur without there ever being a mutation in an area of the DNA coding for something important. This would be a completely unrealistic expectation for random copying mistakes to achieve, because it is far more likely that random mutations will damage the existing DNA than ever develop something new and beneficial like wings.

It is also worth mentioning that there is a critical minimum level that biology itself must reach to form a sustainable whole. Plants rely on animal respiration for carbon dioxide, and yet we rely on plants to make oxygen and convert the sun's energy into sugar. The simplest forms of life (bacteria) rely on the highest forms of life for food, by decaying plant and animal matter. Flowers rely on insects and insects on flowers – which came first? Overall, the extent of complexity interwoven throughout all levels of biological hardware strongly supports the conclusion that life could not have been built up in stepwise fashion. Thus, insight from modern Science proves that *Darwin's own criterion for the falsification of Evolution is well-and-truly exceeded.*

-The Software Problem-

As far as we know, life cannot exist without the biochemical programming contained within the DNA molecule. This is because cellular function and anatomical design is complex, and therefore requires complex programming to build, maintain and replicate. The instructions for these

---

82 See *In Six* Days, Andrew McIntosh, pg 146.

functions come through the deciphering of the information in the DNA, the content of which is determined by the sequence of nucleic acids. Each unit of 3 nucleic acids codes for one of the 20 essential amino acids that biological systems utilize to manufacture proteins, and it is those proteins which perform all the necessary biochemical functions of the cell. Thus, life is dependent on the specific sequence or code within the DNA molecule.

However, because the DNA codes for a polymer of amino acids which must fold correctly to form a protein, it is usually the case that the sequence must be *exactly right* and transcribed without mistakes for the cell to function correctly. There are cases in biology where proteins in different organisms can have a slightly different sequence, so long as the changes do not occur in certain critical areas of the folding structure or active site. Often it is the case, however, that a single point mutation in the DNA can have disastrous consequences when this mistake is transcribed into proteins. For example, biochemical diseases such as sickle cell anemia, phenylketonuria, cystic fibrosis, hemophilia and progeria are caused by single point mutations (one-letter copy errors) in the DNA.[83] The reason why biochemical diseases can be acquired so easily is because the biochemical pathways that the body controls to manufacture or breakdown certain chemicals are irreducible complex chains of reactions, and it only takes one copying error to change the shape of a protein, and the failure of one protein causes catastrophic failure of the entire chain of events. Furthermore, the fact that genetic mutations are accumulating overtime is the opposite of what we would expect from Evolution, and is in fact far more consistent with the concept of a once-perfect genetic language slowly being corrupted over time.

A simple experiment to test if random mutations support Creation or Evolution would be for people to live in close proximity to nuclear radiation, because it is known that this increases the rate of mutations. Of course what we would find is that mutations are harmful to one's health, and we would not see the generation of new, innovative features which increased survivability. Hence radiation is considered harmful, rather than desirable.[84] If the DNA code cannot be altered without it causing harm, this supports the idea that we are starting with a virtually perfect code (allowing for the accumulation of genetic mistakes from creation till now) which is being corrupted. This is obviously the opposite of Evolution – which conceptually involves the addition of new information through mutation to a random genetic sequence, which eventually culminates in the perfection of those genes through survival of the fittest.

When it comes to the origin of life, the problem for naturalists is that there

---

83 Pg 128 (CH8) *The Creation Answers Book*
84 In the case of 'radiation therapy' for cancers, radiation is used because it is harmful and kills living cells (it is more lethal to the rapidly dividing cells of the cancer).

are very low probabilities of arriving at correct coded sequences by chance – depending on how long the sequence is and how many errors you are allowed to have. What we now know, since the discovery of DNA, is that life requires very long, and more often than not, perfect sequences of coded language. This forms an insurmountable problem for Abiogenesis, and simply stated, the odds of any sequence of amino acid necessary for functional proteins coming about by chance are *prohibitively small*. Consider the mathematical probabilities of getting a strip of, say, 10 amino acids in the right specified sequence. The 'right' sequence is one which correctly codes for a protein that performs a certain function. There are 20 essential amino acids that biological cells use to build proteins, so for the first amino acid there would be a 1/20 chance of getting it right, the second 1/20 and so on until all 10 sites were filled. Therefore, there is a $(1/20)^{10}$ or 1/10,240,000,000,000 chance of getting the overall sequence correct by random selection of amino acids. This is a simplified version of the events, because in reality the cell has instructions (start/stop codons) for where to begin and end the transcription. Now let us consider some biological scenarios:

> The one protein example - The simplest naturalistic scenario is that life began with protein polymers. To isolate the problem to just the information content, let us assume that the environment is conducive to polymerization, and that all the cellular machinery necessary for life is in place – including a cell wall (the 'hardware problem'). We'll also work off the concession that all the amino acids have the right chirality (L). If we consider a short protein consisting of only 100 amino acids, using the 20 biological amino acids there would be 20^{100} or 1.27×10^{130} different combinations possible. Now that is a very big number, even compared with generous estimates of the total number of atoms in the Universe (about 10^{80} atoms). Let us now consider the chances of getting the amino acid sequence that forms a functional protein. One might think that there could be billions of 'correct' combinations of amino acids, but this is not the case because proteins that have a biological function have a very specific form and function which, for instance, allows them to catalyse a particular chemical reaction – and this is no simple task! Even if there were a billion functional combinations, there would be 1,000,000,000 chances out of 1.27×10^{130} or 1 chance in 1.27×10^{121} of finding one of those functional sequences by chance. Compare this to the maximum possible number of combinations of atoms in the Universe that could have ever existed: that is, if every atom in the Universe (10^{80} atoms) interacted at the fastest rate possible (10^{12}/s) and for 30 billion years (10^{18}s), and if every one of these interactions produced a new molecule, there could have been only 10^{110} unique molecules that could ever have existed – still far less than the chances of finding one functional protein.[85] Of course, one protein is very far

---

85 John Baumgardner uses this calculation, *In Six Days,* pg 207.

away from being life!

> 75 protein example[86] – Earlier we looked at how the expression of the DNA in living cells requires a factory containing at least 75 protein machines. The problem in this example is that each of the proteins is highly specialized and is interdependent on the other proteins for the overall system to function. This means that the odds of having the right random assembly are very low, because this time the proteins have to be in a very specific sequence. In other words, rather than there being billions of candidate combinations, there would be only 75. But to allow some generosity in the calculations, let us work off the generous assumption that we need to get only half of the sequence right. Let us assume an average length of, say, 200 amino acids and we need to get 100 amino acids in the right sequence. The odds of getting those sequences right, assuming also that the polymers were cut to the exact length, would be $1 / (20^{100})$ x 75 = 20^{7500}, or 1 chance in 3.78 x 10^{9700} - a totally unrealistic possibility. These proteins also need chaperon molecules to assist the folding process, so that the correct 3-dimensional conformation is achieved. Also being ignored in this calculation, is the fact that these proteins need to be in an environment where temperature, pH, salinity and numerous other crucial factors need to be controlled.

> Minimum requirements for life example – Let us now consider the odds of arriving at the protein sequence of an organism that is currently alive. Whilst we can fantasize about different kinds of life evolving, we must be able to explain the kinds of life that we *do* observe. The smallest known self-replicating organism is called *Mycoplasma genitalium* and has only 482 genes (from 580,000 bp).[87] This organism is only a few hundred atoms across, and is a sort of tiny parasite which relies on other larger cells to survive. As such, it cannot survive on its own without more complex life forms, and is below the complexity required for the first living cell. As far as we know, self-sustaining and self-replicating organisms cannot have less than a certain level of complexity, and this can probably be represented, for example, by a simple bacterium like *E. coli*. This has 4,288 genes from 4,639,221 nucleotide bases in the DNA[88]. That's an average protein of about 360 amino acids in length. While some single mistakes can be crucial and others benign, let's allow for an average of 10 mistakes per protein, meaning that we need 350 / 360 in the correct sequence. Again, isolating the information content as assuming all the hardware etc was in place, there would be $1 / (20^{350})$ x 4,288 = $20^{1,500,800}$ or 1 chance in 1.8 x $10^{1,950,000}$ of attaining the right sequences by random chance. Or simply working from the DNA sequence: if we allow nearly 40,000 mistakes,

---

86 John Marcus does a calculation similar to this for *In Six Days*, pg 162-164.
87 A. Goffeau, *Life With 482 Genes Science* 270(5235):445-6, 1995.
88 This example is used by Jerry Bergman for *In Six Days*, pg 16.

and working from the 4 nucleotide bases we get 1 chance out of $4^{4,600,000}$ or $8.8 \times 10^{2,760,000}$ of getting the sequence correct by chance alone. Realistically, there would be some leeway for mistakes to be made in the form of useless proteins being manufactured, but this is restricted because of the possibility of finding a harmful protein that damages the system. The reason why the numbers need to be so restricted to be realistic is because these proteins form functional biological systems, not random collections, and as such they must not only be functional, but also the correct function to fulfil their role in the functioning whole. These calculations do, however, give you an approximation of the sort of odds required for a theoretical first cell. Ultimately, any calculation dealing with the random sequence of proteins, DNA or RNA demonstrates that impossible odds must be overcome for Abiogenesis.

The important thing to note about these examples is the entire sequence must be right solely by random chances. Evolutionary theory asserts that environmental pressures can 'refine' the DNA by natural selection, but this mechanism is irrelevant when it comes to the first theoretical cell. This is because natural selection can only operate through *many successive generations*, and the reality is there are not going to be any offspring of a cell that does not self-replicate. There is a very high technology involved in the replication of even the simplest living cell, and this represents a very high threshold of functioning proteins and correctly sequenced DNA. Therefore, until such 'perfection' is reached, we can expect that the Law of Biogenesis will hold true. It is therefore impossible for a conceptual first cell to acquire the correct computer programming needed for life by chance alone.

Now, if the odds of getting the DNA sequence right by random chance are very low, then we must accept that the chances of the alternative are very high. Organisms such as the *E. coli* mentioned above do exist, and because there is such an infinitesimally small chance of them being *randomly* assembled, the chances of it being *non-randomly* assembled must be astronomically high! The lower the odds for *random* assembly, the higher the odds must be and the greater certainty we can have in *designed assembly*. Put it this way, if someone believes that the DNA code from the simplest self-replicating organism (above) was *not* created by design, then mathematically there is somewhere in the order of a $1 / 10^{2,000,000}$ chance that they are right! Conversely, if someone insists that such an organism is so marvellous that it must have been created by an intelligent designer, by the mathematics of the DNA sequence alone, there would be somewhere in the order of a $(10^{2,000,000} -1) / 10^{2,000,000}$ chance they are right (i.e. certain). Consider this alongside more complicated organisms like ourselves, and alongside the total number of correct DNA sequences that there must be to account for every kind of life in all of nature! Clearly, it is more scientific to believe in a creator God, because it is more consistent with the facts. These examples alone virtually constitute a mathematical proof for God.

An interesting thought experiment is to consider the ruminant stomach of cows. No higher organisms (Eukarotes) have the ability to break down cellulose (B-glycosidic linkages), but some bacteria can. For this reason animals such as the cow have a ruminant stomach where a particular kind of bacteria is cultured, and the bacteria breaks down the grass for the cow. The ability to break down cellulose comes from a single enzyme (in the bacteria) called *cellulase*. Now this is a very curious phenomenon, because if the DNA sequence was the result of random chance, then the chances of evolving this single gene would have to be many times greater than evolving the entire irreducibly complex system of the ruminant stomach! More reasonably, this is a feature characteristic of interdependent design. Animals capable of digesting any plant material could quickly digest whole forests and become the dominant species on Earth, throwing the overall ecosystem out of balance. So it seems more reasonable that the cow is designed as part of an overall ecosystem, rather than just what would be the most advantageous for its survival.

Conclusion - Information

What we see in the DNA is an example of 'information'. Information can be defined as an ordering of matter to convey a message.[89] Matter can never be information, but is the *medium* by which information is stored. So, for example, the information on a CD can be read and turned into music, or it can be transferred to a tape or converted to an MP3. But because the actual information (the music) can be transferred from one median to another, clearly it is not a physical thing in itself. That is, the 'information' is the specific ordering of the sound-waves, the 0's and 1's on the CD, or the magnetic regions of the tape, but it is not the tape itself (or else it could not be transferred). In the same way, information in your brain is not just brain cells, but is the message contained in the ordering of those synaptic connections. Otherwise, information within the brain could not be removed without that part of the brain being removed!

Therefore, information is in fact a kind of *quantity* in itself. There is **matter**, **energy** and then there is **information** - where matter and energy is the medium by which this thing called information is stored. The information/message held in such a medium is therefore not a physical (or energetic) quantity, and therefore cannot be generated by a physical (or energetic) process. Matter cannot self-organize to contain something which is immaterial by definition. Hence, there are no known *physical processes* which generate information. Yet information unmistakable exists, as evidenced from the fact that this very sentence contains a message in addition to the physical ink on the page. Matter can only be

---

89 See Werner Gitt, *In the Beginning Was Information*, 1997. This book explains information theory, and the argument which leads to showing that information ultimately comes from a message sender – God.

ordered by something which itself is immaterial, and the only thing we know is capable of generating new information is the human mind.

The SETI (Search for Extraterrestrial Intelligence) program is an initiative designed to scan the heavens for radio signals from outer space. There are some radio waves coming from outer space, some random noise and some repeating patterns, but despite these, computers are able to sift through these and in theory find anything which contains something more – a coded message! They presume that this would be in an unrecognisable language, but the structure of any information is distinguished from noise, because it is a complex arrangement of matter exhibiting properties of arrangement that does not result from properties of that matter. If we were to find such a signal, then we would quickly conclude there was an intelligent message sender, because we know that only intelligence can order matter to convey a message above the properties of that medium.

This being considered, it is then obvious that DNA is information rich, and because this ordering is not a result of the chemical properties of amino acids, we know that the source of this information must be from an external intelligence. Life today is the copy of the information held in the parent DNA (life comes from life), but when we consider the origin of that life, we must conclude there was a pre-existing, non-physical information parent. If we found a DNA code written into a radio signal from outer space, we would conclude there was intelligence out there to send it.

The software and hardware problems are both unsolvable for Naturalism, and combined they show that Abiogenesis has ridiculously high odds against it. Faith in Abiogenesis violates commonly practised scientific law (the Law of Biogenesis), and involves believing in properties of matter for which there is simply no supporting evidence. The fact of the matter is that life is coded for by DNA, and DNA is ordered to contain information that cannot have arisen from *any* physical process. Therefore, it is a logical certainty that the information contained within the fabric of life itself originated from an external intelligence.

Chapter 4 - Cosmology

Whilst the events of the past will always remain in the realms of calculated guesses, there are some things we can know for sure about the Universe in which we now live in. One of those is that the Universe must have had a beginning:[90]

The first reason for this is that the Laws of Thermodynamics[91] necessitate it. The First Law of Thermodynamics states that energy can neither be created nor destroyed, but only transformed from one form into another. By considering the Universe as a giant closed system (where there are no outside influences), the First Law indicates that the overall amount of energy in the Universe must have always been constant. Thus, all the energy in the Universe could not have been generated from within the Universe, and since energy does exist within, it must have been input from *outside* at the beginning of time. The Second Law of Thermodynamics states that matter always tends towards the (more stable) lowest possible energy state (where disorder/entropy is the highest). Thus, over time the amount of *available* energy that can be used is always decreasing. Therefore, in an infinite amount of time all the energy in the Universe would have long since reached a state of equilibrium where no part of the Universe could have more energy than another (thus, no available energy transfers would be possible). We have not come close to reaching this point, and so the Universe we find today cannot be infinitely old.

The second reason why the Universe must have had a beginning is because 'infinity' is a physical impossibility (infinity is a mathematical concept, but actual infinite sets cannot exist in reality). It would be impossible for the Universe to be infinitely old, because that would mean there has been an infinite amount of time that has passed for us to arrive at the present moment. This cannot be the case, because it is impossible to traverse an infinite amount of time by the definition of 'infinity'. Infinity is a place you can never reach! For example, if today was time=infinity, then yesterday would be infinity minus one, and tomorrow would be infinity plus one – an impossibility since infinity cannot be added to. Likewise, there cannot have been an infinite number of events leading up to the present time, because each new event that occurs again would be adding to an infinite number. Given that there is an increasing flow of events or time continually adding to today's, we can be certain there are not infinite numbers of these, because it is impossible to construct an infinite time-line without either end. Therefore, time itself must have had a beginning along with the beginning of all events (matter and time began).

---

90 These two arguments are also used by William Lane Craig in *The Case for a Creator* by Lee Strobel, chapter 5.

91 These Thermodynamics Laws are some of the most certain in Science. For example, they are critical in understanding Chemistry - all matter tends towards higher entropy (disorder), and this drives chemical reactions towards the more stable compounds.

This leaves us with two possible scenarios:
1. The Universe is self-caused by natural law.
2. There is an outside cause (a creator).

Self-Caused: The Big Bang

The naturalistic/self-caused explanation of the Universe is perhaps best typified by the Big Bang model, which is widely accepted as being the explanation fitting most closely to the facts (from a self-caused perspective). There is a general consensus amongst the scientific community that the Universe is expanding, so at some point in the regresses of the past, the Universe must have been smaller. It is observed from Earth that the galaxies which are farthest away are the most red-shifted, and so it is thought that these are receding from us the quickest (according to the Doppler-effect explanation[92]). It is thought that by extrapolating back through time we would get to a point where all matter in the Universe would be in roughly the same region. If this was the case, then the gravity of such a region would be so large that we would expect it to be pulled into a single region, often pictured as a point of singularity or atomic collapse (a sort of primordial black hole). Thus, The Big Bang is a model that cosmologists use to explain the history of the Universe, where essentially the Universe originated from a point of dense matter and/or energy which expanded outwards creating space, time and matter as we know it today.

The Big Bang Model[93]

Below are the 5 major aspects that the Big Bang model describes as the history of the Universe.

(1) Singularity - Before time began there was a point of 'singularity' which was a uniform region of all matter and energy. The rapid expansion of this infinitely dense and hot region is the source of all matter we now observe, along with space and time itself.

☒ Singularities (like the inside of a black hole) are thermodynamic dead-ends. They contain an extremely stable, super-dense material (where all the atoms have collapsed), and cannot be converted back into other forms of matter. Therefore, it is physical impossibility for the matter we have today to have originated from a singularity. This is a major problem for big-bangers, because if all the matter of the Universe was

---

92 The Doppler Effect is the most obvious when the sound of a ambulance siren increases as it approaches, then decreases as it travels away. The speed of the object generating the sound affects the sound waves you are hearing. The same is thought to be true for light from distant galaxies.

93 See *Dismantling the Big Bang* by Alexander Williams and John Hartnett (2005), who use Joseph Silk's book as a reference for what is generally accepted among cosmologists about the Big Bang -*The Big Bang*, 3rd Edition (New York: Freeman & Co., 2001).

in one place, it would condense into singularity and never return from it! This is much like what we expect to find inside a 'black-hole', which is considered to be the end of matter - where nothing (even light) escapes. Therefore, as far as we can tell from physical laws in operation today, it is impossible for the Universe to have originated from this 'black-hole' state.

(2) Inflation - After a very short period of expansion (the Plank time of 10^{-43}s) pure energy is converted into subatomic particles.

⊠ It is thought that the Universe must either be in a state of expansion or contraction, because a neutral speed would be too precise to achieve by chance from a rapid initial expansion. If the Universe did not expand at a high enough rate, then it would certainly collapse back into a gigantic black hole (The 'Big Crunch'). If the Universe expanded too quickly, then it would continue expanding forever so quickly that solid objects would never form (The 'Big Chill'). One of the biggest problems for Big-Bangers is that the initial expansion of the Universe would have to be at *precisely* the correct speed, because otherwise one of those two extremes would result. The Universe as it is would require a 'big-bang' that was extremely 'fine-tuned' at the right expansion rate, perhaps to the precision of one part in 10^{120}. This seems far too precise for a random 'bang'. [94]

⊠ The expansion of space and time would also need to be perfectly uniform, because otherwise density fluctuations would cause gravitationally driven re-collapse shortly after the initial expansion (forming a gigantic black hole).[95] Therefore, the likelihood of an inflation rate being so perfectly right as to result in a universe are so astronomically small, it can be ruled out as happening by chance alone.

(3) Matter from Energy - Normal atoms of Hydrogen and Helium begin to form over the next 100,000 years as the expanding region begins to cool. After about 300,000 years the expanding cloud of (mostly) Hydrogen gas becomes transparent and the energy causing the expansion begins to subside. The intense glow of the initial radiation continues as Cosmic Microwave Background radiation.

⊠ When matter is produced from energy, Quantum Law predicts that particles can only be formed in quantum pairs, and therefore for every particle of matter there must also be an equal and opposite *anti-matter particle*. Should these two particles again meet, they would annihilate each other and the original energy would be returned. However, our universe is made of ordinary particles of matter, not 50/50 matter and antimatter, and therefore our Universe could not have been produced

---

94 *Dismantling the Big Bang,* pg 123.
95 *Dismantling the Big Bang,* pg 121.

by the conversion of energy into matter through quantum pair production. The implication of this is that the matter in the Universe must have always existed in some form, and as already explained, this could not have been from a singularity. Thermodynamics means that matter cannot have existed forever, and therefore there must have been a non-material cause for matter in the finite past.

(4) Galaxies and Stars - About 1 billion years after the beginning, galaxies and stars began to form by gravitational collapse. The temperature and pressure generated inside these balls of gas initiate fusion reactions - which then acts as a fuel.

☒ Up to this point all that has been produced by the Big Bang is an expanding cloud of gas, and there really are no good explanations for how this began to contract in localized regions for solid objects to eventually form. In order for objects to begin forming, expansion would have to decrease somewhat making way for localized contraction. However, this is a very big mystery because a contracting gas cloud of this mass would collapse back into the singularity (a dead end). Therefore, there would need to be a very, very precise balance between expansion and gravity, and the Universe would need to be very uneven for gravity to bring the gas together in localized regions. This could not have been the case, because the original expansion itself necessitated a *perfectly uniform* distribution of matter to avoid re-collapse into singularity shortly after expansion began!

☒ Additionally, if it were true that there were density irregularities in the matter of the early Universe which worked against the expansion and generated the formation of objects, then we still would not expect there to be the formation of stars and galaxies. A collapsing mass the size of a star would form into a black hole, and any galactic sized mass of gas would even more certainly collapse into a super massive black hole. "The universe is, by definition, the planets, stars, and galaxies that surround us. Insofar as big-bang theory does not explain the origin of these objects, then we can say that big-bang theory *does not even address the question* of the origin of the universe."[96]

(5) Stellar Evolution - The first generation of stars were thought to have been relatively unstable and exploded to produce metallic stardust (heavier elements are thought to have been produced inside stars by fusion reactions), which eventually went on to form the stars which we now have. Our own solar system is thought to have formed from a nebula of gas which condensed into the orbiting bodies we now observe (including Earth).

---

96 Taken from *Dismantling the Big Bang*, pg 129.

- There would have to be a large number of these exploded stars to account for the formation (and heavier metal content) of 'modern' stars, but a large number of exploded stars means a large number of black holes – which has not yet been observed. If there were a large number of black holes around, as would be expected if there had been many generations of stars before now, then they would also capture much of the matter needed to form new stars.

- Stars are made up of glowing balls of gas which are held together *by gravity*, yet gravity is not strong enough a force to form stars from a gas. Gravity would need to overcome the repulsive force a gas exerts when it is squeezed together (the higher the pressure, the more it tries to escape). It is proposed that fluctuations in the density would overcome this problem, but these fluctuations would need to be exact – too small and no star would result, too big and a black hole would be the result. The problem is that these density fluctuations in turn would have to originate from other solid objects ie. shock waves from other exploding stars. As far as we know you need stars to produce stars!

- Fusion cannot produce elements heavier than iron, because after iron the nuclear stability decreases and nuclei tend to undergo *fission* back to iron. The abundance of heavier elements is unexplained through the Big Bang.

- There are several lines of evidence that stand against the idea that our solar system came from a single nebula or dust cloud. It is not known how dust granules could stick together in large enough numbers to form self-gravitating bodies, and this is called the 'sticking problem'. Each of the planets appears to also be made of very different kinds of material. There is no conservation of angular momentum either. The sun, for example, which comprises some > 99% of the mass of the solar system, possesses only 2% of its angular momentum.[97] Therefore, they cannot have originated from the same spinning object. Further support for this comes from the fact that the planet Venus spins in the opposite (retrograde) direction from the other planets, and therefore again could not have come from the same spinning object.

Summary – The Big Bang model cannot explain how the Universe formed from a singularity, or what caused such an enormous expansion, where matter came from, what stopped the expansion so precisely and why there were localized regions of contraction, how galaxies and stars are formed by gravity alone, or how our solar system formed. Each one of the five major aspects of the Big Bang model is unrealistic, some violating well-known scientific laws. Taken together, it is hard to see what the Big Bang model actually does explain! The Big Bang is a picture of the

---

97 *Refuting Compromise*, pg 169.

Universe's history, but it cannot be considered a scientific theory by any stretch of the imagination.

Better Science

The Big Bang is a guessed reconstruction of past events based on limited observations in the present, and is only *one model* by which scientists attempt to explain the origin and formation of the Universe (Naturalistically). Alternatives to the Big Bang exist. Although the same evidence is used to formulate different models, the important difference is the *starting assumptions*. When these are challenged, we shall see that there are scientific models that better explain the observations we have today.

The model being suggested here, is one where we perceive the Universe as having been intelligently ordered, and which has expanded from a much smaller original state. This approach relies on much the same evidence which fuels development of the Big Bang, but in most cases has different starting assumptions.

(1) The Galactocentric Universe - From Earth's position of observation it appears as if space is uniform in all directions. There are two possible explanations for why this might be: (1) The Universe is infinite in all directions with an infinite number of galaxies and stars, and thus from any point in space the observations will be the same (the 'Cosmological Principle'). This is called an 'unbounded' Universe and implies that if you kept travelling through space you would never reach a point where you run out of matter to observe ahead of you. (2) The second option is that the Universe is a spherical shell of matter of limited size with our galaxy being near the center.

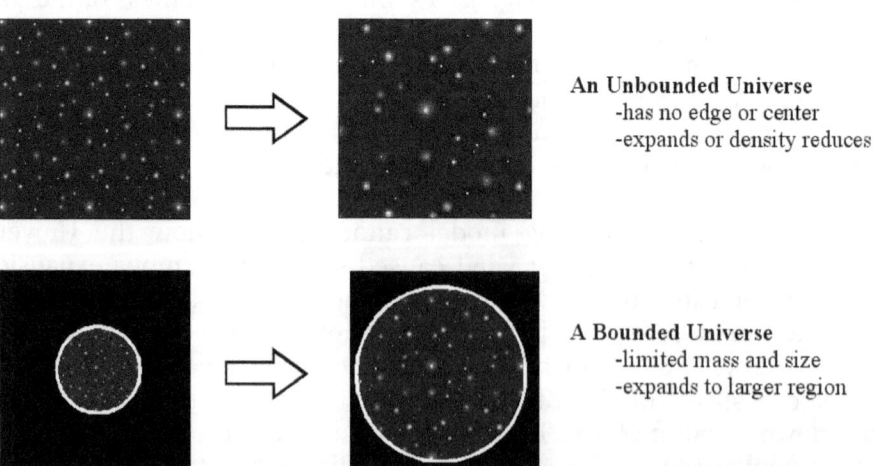

An Unbounded Universe
-has no edge or center
-expands or density reduces

A Bounded Universe
-limited mass and size
-expands to larger region

☑ Many cosmologists tend to favour the first assumption (an unbounded universe), because the second seems to imply that there is something strangely special about planet Earth (which would appear to be in the center) - and this is a disturbing picture for many. Also, under the first assumption one does not have to consider time dilation factors, because a universe without center and without edge does not have a center of gravity.

☑ The Universe must have an edge (a point where you could get to beyond which there would be no more matter), because there must be a limited amount of matter in the Universe. There cannot be an infinite amount of matter, because if there was, matter would inhabit every part of space (even an infinite space). Neither could the Universe have expanded into an infinite region in a finite period of time – so it could not have started from a limited size initially. Also, if it was infinitely spread out, there would be an infinite amount of free space (which is not the case). It is commonly theorized that the Big Bang did not involve matter coming out of a state of singularity, but rather, the matter of the Universe already existed, and as space itself expanded, matter reduced in density. However, an infinite amount of matter could not have fitted into a finite region, and an infinite region cannot expand to be a larger one (than infinite!). Therefore it makes more sense, logically, to accept that the Universe has a center and an edge.

☑ Quantized Red-shifts - It has been known for a long time that the red-shifting of light coming from distant galaxies is in fact grouped around distinct (quantized) wavelengths. The quantization of red-shifts (the change in colour of light coming from distant stars) therefore suggests that these galaxies are grouped at certain *discrete distances* away from our viewing perspective. In other words, it now appears as if galaxies are not randomly distributed, but are 'organized' into concentric shells of galaxies. This type of pattern could not be observed unless our galaxy was at or near the center of the Universe.

☑ Both the polarization of light coming from distant galaxies and the anisotropy (variations) of the CMB radiation is consistent with the Universe having a north and south pole, and an equatorial region. This is again consistent with the Universe generally rotating around a central point to which our galaxy is quite close.

☑ Also, the most recent large scale map of all galaxies (known as the Sloan Digital Sky Survey) has created a map of the stars, which indicates that there are concentric shells of galaxies centered around our location in the cosmos with a decreasing density the farther out you go from here. This is the opposite of what you would expect from the Big Bang, and conflicts with the Cosmological Principle (an unbounded universe). That is, the evidence suggests that we do inhabit a privileged position in a (bounded) Universe with a center and edge.

(2) Galactic Red-Shifts - It has been observed that many galaxies outside our own have their light 'red-shifted' in proportion to how far away from the Earth they are thought to be. It was initially thought this was caused by the Doppler effect alone, which meant that the wavelength of the light travelling here was altered by the speed at which the galaxies were travelling away from our observation point. If we say that the farther away a galaxy is, the faster it is moving away, then when we extrapolate back in time we would end up with all matter converging to roughly one point in space. Thus, red shifting is one of the major observations leading scientists to the Big Bang model, because it appears that the Universe is (or has been) in a state of expansion. More recently however, people have begun to suggest that this red-shifting in wavelength is mostly caused by the *expansion of space itself* (the 'Hubble Flow') – which also carries these objects along with it. Under this interpretation, the red-shifts themselves are due to the expansion of space while the photons are in transit through it.

☑ Just because the Universe is expanding (or has expanded), it does not follow that it started from a point of *zero size*. A Universe which began in a smaller state would still have the same observations. Seeing the expansion of space itself as the major contributor to red-shifts, this would mean that rather than tracing all matter back to one point, we could in theory trace it back to a Universe *proportionally smaller* than the one we have today. Red-shifting is most likely to be an artefact of the distance that the galaxy *has travelled* as space has expanded. The more red-shifted, the more that the space between us has expanded while the photons have been in transit, and thus their wavelengths have been 'stretched out' proportionally. Therefore, red-shifting can be explained by a Universe which was proportionally smaller, because the outer galaxies would have travelled further to now be at their current locations.

(3) Cosmic Microwave Background Radiation (CMB) - An electromagnetic wave has been discovered to originate from every point of empty space at which a telescope can be pointed. This energy, in the microwave region of the electromagnetic spectrum, corresponds to a (black-body radiator) object of about 2.7K (-270.3°C). This is commonly thought to be the resounding ring left over from the 'sound' of the Big Bang, or the afterglow of the heat released soon after the Universe began expanding.

☑ There is not enough time for the energy from the initial (Big Bang) expansion to have travelled to all parts of the Universe and evened out. This is called the Horizon Problem. Under the Big Bang model, there must have been some variation in the expanding mass and energy of the Universe at some point to eventually form stars and galaxies. However, the background energy we observe today from the CMB is very uniform. Therefore, the problem for Big Bangers is that there has not been enough time to allow the light from hot and cold regions of the Universe to become as uniform as it appears today.

☑ All we really know is that this light exists, but we cannot know for sure what the source of it is. Some believe that the background radiation we observe is nothing more than the result of all parts of space being bathed in distant starlight, and is simply the temperature of empty space. Others believe that it is caused by the thermalization of starlight with different molecules in space.[98]

☑ CMB can be better explained by White Hole Cosmology and Relativity (to follow in section 5). Under this model, the Universe is accepted to have expanded rapidly outwards. This would have appeared as an enormous ball of expanding light, because the force was strong enough to repel photons outwards (the opposite of a black hole). The same stretching out of the wavelengths of visible light would have the same effects on the thermal radiation from the early Universe. That is, the original light present in the early Universe is 'red shifted' to the microwave region of the electromagnetic spectrum. General relativity also allows for enough time for this light to have become uniform and reach all parts of the Universe.[99]

(4) Observations of A Young Universe – We are commonly informed that the Universe started with a bang somewhere in the order of 15 billion years ago. This, is of, course assuming some type of Big Bang scenario. However, there are a number of actual observations from the Universe that tend to suggest otherwise – a young Universe.[100]

☑ Missing SNRs – It is estimated (from observations) that, in galaxies like our own, every 25 years a star should explode (a supernova) and form an expanding cloud of gas (a Supernova Remnant – SNR). It is estimated that there should be at least 7000 of these in our own galaxy - by the billions of years' time-scale. At present, only 200 SNRs have been observed, which is consistent with only about 7000 years of cosmic evolution.[101]

☑ Spiral Galaxies – Many galaxies have a distinct spiral structure which is caused by the differential rotation of stars about a central axis. This pattern is even observed in our own Milky Way galaxy where the inner stars rotate at a higher speed than the outer ones. Although our galaxy

---

98 *Dismantling the Big Bang*, pg 127, 284.
99 *Starlight and Time*, pg 122-125.
100 See *Evidence for a Young World*, *Refuting Compromise*, Ch 11, *Dismantling the Big Bang*, Ch 5, and *The Creation Answers, Book* Ch 5.
101 *Refuting Compromise*, pg 349.

is said to be 10 billion years old, it is thought that after only a few hundred million years it would lose the spiral shape and become a featureless disc of stars. It is believed that it also takes a certain amount of time for the spiral shape to form, and therefore very young galaxies should not yet have acquired them. This implies that these galaxies are the same age as our own, and therefore we should reconsider how long it has taken their light to reach this far.

☑ Galaxy Observation – If we are really seeing galaxies that are billions of light years away, then it follows (in the Big Bang scenario) that we are seeing things the way they were that length of time ago. That is, as we look farther out into space, we should be 'looking back in time' and should observe younger and younger galaxies. The farthest galaxies (say 15 billion light years away) ought to have been formed not that long after the beginning of the Big Bang, and should appear *very different* to the closest ones if their light really took billions of 'years' to reach Earth. However, galaxies are very uniform in appearance, regardless of distance, and do not differ that much from our own 'mature' galaxy.

☑ Comets – Comets are supposed to be about 5 billion years old – the same age as the solar system. However, each time they pass by the sun they lose a lot of their material. It is thought that no such comet could survive longer than about 100,000 years, with an average age of most comets being less than 10,000 years.

☑ The Moon – The Moon is receding away from the Earth at an estimated speed of 4cm/yr. This is thought to be because each year the ocean tides cause a certain amount of friction which slows the Earth's rotation. As a result, the angular momentum lost through Earth's rotation is gained by the Moon, thereby increasing its speed and orbit height. Obviously then, in the past the Moon must have been closer to the Earth than it is now. Now, the moon could only be so close before it enters the Roche limit (about 18,400 km out, 460 million years ago at current speed), where gravitational forces would tear it apart. At its current speed it would have been touching the Earth only 1.4 billion years ago.

☑ Superclusters – It has been found that many galaxies exist in large groups called 'Superclusters', which can contain as many as 1000 galaxies each. If we assume the expansion of space over time, then such Superclusters would have expanded and should no longer be grouped as they are. However, it appears as if we are viewing them as if they are very young (less than a billion years).[102]

☑ There are also observations of Saturn's rings (and recently discovered rings from other outer planets) consistent with a much younger solar

---

102 *Dismantling the Big Bang*, pg 200.

system. Also, the magnetic fields of other planets in our solar system are most likely to be freely decaying (like our own), and hence it is more realistic to suggest a younger, rather than older, age for these.

(5) Light from Distant Galaxies - One of the arguments used to support the Big Bang (or more often against a recent creation) is that we can observe light from distant stars, which, because of the huge distance they are away, must have taken millions of years to reach us. Thus, they argue, the Universe is millions of years old.

☑ The Horizon Problem – The Big Bang model also has a starlight and time problem of its own, because it cannot account for the uniformity of the CMB (section 3 above). Since there has not been enough time for the CMB radiation to have spread to all parts of the Universe and evened out, the light travel problem is not solved by reverting to an old Universe and Big Bang scenario.

☑ White Hole Cosmology – If we work within the assumption of a Universe that has expanded, but then we also assume that there is an edge and center (a bounded universe), then we can arrive at *White Hole Cosmology*. This is the opposite of a Black Hole - where matter is pulled by gravity towards a point as the 'hole' expands. The size of the hole is proportional to the mass within it, and is black because within a certain radius (the Schwarzchild Radius, or Event Horizon) the light cannot escape (the escape velocity is greater than the speed of light). In White Hole Cosmology, an originally small Universe is depicted as expanding outwards as the radius of the expanding region (the Event Horizon) shrinks and dissipates. Matter is forced out of the white hole region, causing the central mass within it to decrease as it loses more material to the expansion. At the same time, the strength of the expanding force and the radius of the expanding region decreases, eventually reaching zero. Thus, the outermost galaxies travel the farthest, and the innermost galaxies travel the least - so that the original proportions are maintained. Rather than galaxies travelling through space, it is space itself that expands and carries the galaxies along with it. This is thought to be the origin of the red-shifting (stretching out) of the wavelengths of light - particularly from the galaxies which are more distant from the center.

☑ As a body passes through the event horizon (the edge of the white hole), the expanding force reduces and significant time dilation (relative to other localities) occurs as a result. According to Einstein's theory of General Relativity, as an astronaut was sucked into a black hole they would observe time elsewhere begin to speed up (as time for them slowed down), and they would see clocks going much faster and events occurring much quicker (shortly before they would be crushed smaller than an ant). Conversely, travelling out of a white hole looking back in, an astronaut would see time slowing down on those clocks, and events occurring much slower than usual inside the white hole. An observer of the astronaut from the inside would be surprised to see time begin to go exponentially faster for the astronaut as they receded into space (although the light from them would not be able to get back into the white hole to observe this). The watch on the astronaut's wrist would be ticking at an enormous speed, and his arm waving goodbye would appear only as a blur. Although he was at a constant 'speed' (distance/*time*), he would appear to exponentially accelerate as time speeded up *relative* to inside the white hole.

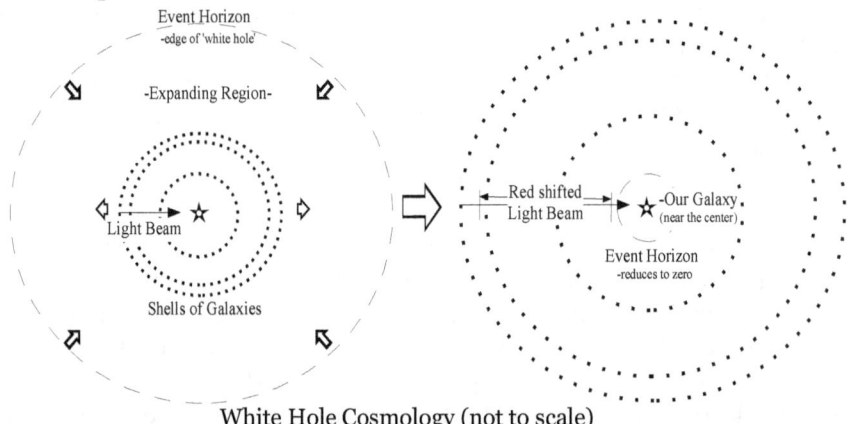

White Hole Cosmology (not to scale)

The result of this time dilation is that long periods of 'time' can pass by in other areas relative to an observer near an event horizon. Now, if the Earth is somewhere near the center of the Universe, then in this model we would have been close to last to travel through the reducing event horizon (which shrinks to nothing around a central point of the Universe), and thus a huge amount of 'time' would have passed farther out in the Universe relative to us. Under this expanded Universe model, light from distant galaxies would have had plenty of 'time' to traverse the huge distance, and as an artefact of the expansion of space they travel through, we would see the light red-shifted. This scenario implies a young (according to our time scale), structured Universe that has expanded into what we now have, and accounts for the starlight and time problem. It is also consistent with the major observations of a galactocentric Universe, red-shifting of light from distant galaxies, and the CMB. Thus, these can no longer be regarded as evidence for the Big Bang. It also avoids the major difficulties associated with the Big Bang (such as an initial singularity and the missing anti-matter etc.).

It also helps to resolve the appearance of young Superclusters. The problem is that these appear to be billions of light years away, and so, under the Big Bang scenario, people understand viewing objects this far away as essentially 'looking back in time' to see the Universe as it was billions of years ago. However, once we consider time dilation factors, we can acknowledge that these superclusters are billions of light years away from us in distance, but we are seeing them as if they were a few thousand years old (as measured by Earth's clocks). As a thought experiment, consider what would happen if the stars were initially created close to the Earth, and therefore visible on Earth from the first day they were formed. As the Universe expanded as a white hole, time ticked relatively faster for them as these stars receded billions of light years away in *distance*. Passing through the Event Horizon, in a short time by our standards, billions of years' worth of light poured out of the stars in proportion to how far they travelled, so that their light filled the space they traversed through creating a continuous stream of light in transit towards Earth. The net effect of this is that the light that we are seeing today is perhaps only a few thousand years old (by our clocks), with billions of light year's worth of photons already in transit. It is then reasonable to say that the world is thousands of years old, and the stars *appear* as the same age as the Earth, even though they are millions of light years away (and millions of years old relative to us).

(6) Fine tuning – Any casual observer of the Universe soon discovers that Earth is uniquely suitable for life. In fact, there are a number of parameters and unexplained coincidences that have such a very low probability of occurrence, that it suggests the known Universe cannot be the result of random chance, or the result of some sort of explosion where matter was randomly assorted.

☑ Existence of Order – order is written into the fabric of the Universe, and it is there for us to discover. For example, we have discovered that atoms and energy move and transform according to precise mathematical principles, but there is no reason why this must be so. Matter is not a thinking or rational entity, and therefore it cannot self-order (and in fact tends towards disorder), but yet we find it to be *precisely* ordered and rationally understandable. We did not invent mathematics, but by studying the Universe, we humans have discovered principles that govern matter. Order appears to be an entity that exists beyond and before matter. Mathematics, together with other guiding principles, is therefore the inherent rationale that we discover within a non-rational (material) universe, and strongly suggests that matter itself has an ordered/rational origin outside itself. One famous scientist is quoted as saying *"A common sense interpretation of the facts suggests that a superintellect has monkeyed with the physics, as well as chemistry and biology..."*[103]

☑ Fundamental Constants – *"Over the past thirty years or so, scientists have discovered that just about everything about the basic structure of the Universe is balanced on a razor's edge for life to exist. The*

---

103 Sir Fred Hoyle, *"The Universe: Past and Present Reflections" Engineering and Science* (1981).

coincidences are far too fantastic to attribute this to mere chance..."[104]
For example, if there were minute changes in the force of either gravity, the strong nuclear force, electromagnetism, or the relative mass of neutrons and protons, life in the Universe could not be possible. *"In light of the infinitesimal odds of getting all the right dial settings for the constants of physics, the forces of nature, and other physical laws and principles necessary for life, it seems fruitless to try to explain away all of this fine-tuning as merely the product of random happenstance."*[105]

☑ Planet Earth – Planet Earth possesses a number of precise characteristics which enable life to thrive here.[106] For example: of the different kinds of galaxies that we can observe, our type is the safest, and our position in this galaxy happens to be the safest (it is also ideal for observing other galaxies). Earth is in a particularly perfect obit around a star with just the right size and spectral luminosity to have temperatures suitable for life here on Earth. We are just the right distance away from the Sun, and the speed of our rotation is just right to regulate the temperature accordingly. We have Saturn, Uranus and Jupiter outside us which draw comets away from us, and we have Venus and Mars which protect us from asteroids (there is an asteroid belt between Mars and Jupiter). The tilt of the Earth of 23.5 degrees provides seasons that regulate the temperature. We have a magnetic field (thought to be generated by the movement of liquid iron inside the Earth's core) that shields us from cosmic radiation by deflecting it towards the polar regions. We have a moon that contributes significantly to the stasis of the climate by stabilizing the Earth's tilt, and contributing towards large-scale oceanic circulation. It just happens to be the right size and distance from us. The size of the Earth is big enough to retain an atmosphere, while being not too big to smooth out the Earth's crust (which would result in the planet being engulfed in water). The atmosphere has just the right mix of gases for life, and allows the right amount of energy to penetrate to the surface while filtering out harmful radiation (ozone). This unique convergence of precise parameters that are necessary for life to exist is leading some scientists to conclude that Earth is the only planet in the entire Universe that has any possibility for life.

Summary – Cosmology involves a lot of calculated guess work based on limited evidence. When weighed, this evidence clearly favours models based on a young universe that has expanded. Gravitational time dilation is a well-known science based on Einstein's equations, and explains the anomaly of being able to see distance stars in the present day, along with

---

104 Interview with Robbin Collins in Lee Strobel, *The Case for a Creator*, pg 131.
105 Lee Strobel *The Case for a Creator*, pg 135.
106 These details are explained by Guillermo Gonzalez & Jay Wesley Richards in their interview with Lee Strobel in *The Case for a Creator*.

accounting for red-shifting and CMB. The Universe we observe today also strongly suggests design, and it would be unreasonable to attempt to account for it by chance alone.

Conclusion: Created and 'Stretched Out'

The Universe is Caused

It has been shown that a model of a self-caused Universe is utterly impossible. A number of scientific constraints prevent any Big-Bang type scenario, and besides, everyone knows that objects remain in a state of rest until an external force is applied to it (Newton's law). Everything that begins to exist owes its existence to something outside itself, and it seems inevitable that since the Universe began to exist, we need to invoke some sort of outside agent as a first cause.

Since the Universe needs an external cause, it should not be surprising that we also find examples of order at every level of reality. Firstly, we can say that order, such as mathematical guiding principles, exists in a material which cannot self-order. This suggests a pre-eminence of mathematics originating outside of the (non-rational) physical creation. Secondly, we can see that the interactions between such fundamental levels of order are in themselves precisely tuned for the existence of anything (different forces are precisely calibrated with each other). Reality itself seems to be an incredibly delicately balanced system, with a very, very low probability of existence. And lastly, the planet we live on is again a convergence of a large number of variables set at just the right level to be conducive to a habitable environment.

We can say then, with good confidence, there must be some sort of creation event, involving an external (outside the Universe) cause, which we can also attribute as the origin of the order we find in it. Perhaps we can therefore make some tentative conclusions about what this cause might be: it must be external to the material/natural Universe, and therefore it must be immaterial or supernatural. Since this entity existed before time and space, it is outside of time and eternal (existing in all points of time - omnipresent). We must also conclude that the first cause is the ultimate cause of all other causes, and is therefore greater than the sum total of all energy within the Universe (omnipotent). Since the Universe shows elaborate design, one way or the other, there must be an explanation for such order. We must use a hypothesis that is the most reasonable explanation to account for all the order that we see, and it seems inevitable that intelligence must be invoked as that best explanation (omniscient). For example, a random 'bang' of any sort will not produce ordered systems such as our solar system.

Since the odds of the Universe existing, and planet Earth being conducive to life, by chance, are astonishingly small, we therefore have to conclude that the chances of it being *designed* are astronomically high. Earth, together with everything else in the known Universe, at least gives the *appearance* of being elaborately designed to accommodate the fragility of life, and this is clearly the best explanation over other alternatives. Now, some might argue that the Universe only has the appearance of design, because if it *wasn't* perfectly suited to life, then we wouldn't be alive to be making the observation that it was! We, for example, could be the one success story out of a billion life-forms which never existed, because their world was less suitable for life. Hence, life appears designed for those one-in-a-billion life-forms who do exist. However, just because we are alive to make that observation, it does not provide a reasonable explanation *for* that observation. Thus, we still need to come up with an explanation for why such an unlikely Universe exists in the first place, whether or not we are here to observe it. For example, if the Universe did come about by chance, we would not expect to find it impossibly fragile and incredibly unlikely. We would expect to find a Universe that was so simple it was inevitable that it would happen by chance, with perhaps many near misses for life on planets elsewhere. We would expect to see a world that was *likely* from random, non-ordered beginnings. This is the opposite of what we do find, where life seems so desperately *unlikely* that it suggests some sort of miracle to even Atheistic scientists.

A Biblical Creation?

In the wake of the Big Bang, there is growing scientific evidence to suggest that we live in a galactocentric Universe (with our galaxy at the center of the Universe) that was once smaller than what we presently observe. This fits in better with the observations from the red-shifting of stars and the CMB radiation. If we start with the assumption of a bounded Universe (rather than infinite), then we realistically need to consider time-dilation factors in an expanding Universe scenario, and a White Hole Cosmology emerges. This reconciles how we can see light from distant galaxies in the present day with the observation that many of the Universe's features appear to be very young (on a cosmic time-scale).

What is most interesting about this scenario is that there is a strong suggestion in the Bible that such a thing actually happened! Together with the creation story depicted in Genesis one, there are also a large number of verses scattered throughout the Old Testament that suggest that God at some point 'stretched out' or 'expanded' the heavens. These are curious statements for the Bible to make, and it could be assumed to be 'just poetry', except for the fact that the Bible mentions this about 17 times by at least 7 different authors[107], and it seems to be referring to the physical Universe. Perhaps, this is a divine hint leading us to better understand the nature of the Universe we live in.

---

107 See 2 Sam 22:10, Job 9:8, 26:7, 37:18, Psalm 18:9, 104:2, 144:5, Isaiah 40:22, 42:5, 44:24, 45:12, 48:13, 51:13, Jeremiah 10:12, 51:15, Ezekiel 1:22, Zechariah 12:1.

Chapter 5 – The Age of the Earth

When it comes to the age of the Earth, we are often given precise dates of historical events extending millions of years into the past. However, it may come as a surprise to many that the world *cannot actually be dated directly* by any scientific measurements. We can only take measurements from physical processes that exist *in the present*, and guess how long they have been in operation for. With such an approach, we must employ a certain number of assumptions to come up with historical dates. For time-difference to be inferred, we must first assume that the rate of any physical process we see happening today has been constant throughout history, and that we have a good idea what the initial conditions were like. These assumptions in turn are based on our *preconceived* view of history.

Two Assumptions

There are two basic (and mutually exclusive) geological interpretations we can make about the history of the world - Uniformitarianism and Catastrophism. The assumption used by the prevailing (Evolutionary) interpretation of Geology is Uniformitarianism, and this assumes that the processes we see happening today are responsible for shaping the Earth in the past. It assumes an essentially constant rate for physical processes over billions of years, and as a result, it allows us to estimate the ages of things based directly on their *current speeds*. Catastrophism on the other hand, in addition to current processes, invokes the idea that there must also have been *exceptional circumstances* in the past to explain what we now see in the present. Unfortunately, under this assumption, many dating methods would be invalidated, because with changing speeds and catastrophic events, we cannot make an accurate estimation of history using the speed of current processes.

Under the Uniformitarianism scenario, it seems as though the world must have been around for billions of years. At today's very slow and gradual geological movements, it would take an incredibly long time for the Earth to be shaped into its current topography. However, the same could be achieved in a very short period of time, given a large enough catastrophe. Catastrophism usually employs *Flood Geology* – estimating the Earth's history based on a deluge from huge amounts of water. These two geological scenarios happen to be in direct contrast, because they project significantly different pictures of Earth's history. One cannot hold to both slow, gradual changes and, massive, sudden changes to explain the same data. Consider, for example, how a valley or canyon might have been shaped by water: either the river has been flowing near enough to its current path for millions of years (a little water and a lot of time), or the canyon has been carved out by a massive event leaving the river at the bottom (a lot of water in a little time).

If Uniformitarianism were true, then there must never have been a large-scale catastrophe, for this event(s) would override slow processes and become the *dominating force* that has generated most of the Earth's features. It would then be impossible to use current, 'unchanging' processes to extrapolate back in time to generate an accurate picture of history. Whereas, if Catastrophism were true, then the slow and gradual processes we see today are all succeeded by massive geological movements that are no longer in operation. Once we allow major catastrophes in our model of what has happened in history, estimation of Earth's projected time-scale dramatically reduces, because no longer do small processes have to be happening for aeons of time to have a large-scale effect. Large-scale observations can be explained by rapid, large-scale events. In this scenario, most of the features of the geological column are the result of large-scale aqueous events, with current processes being only residual from what has happened previously. Under this assumption, there is no basis upon which we can formulate a slow and gradual history of the Earth, and therefore no basis to project a history spanning billions of years. Simply put, if there ever was a large-scale flood that shaped the Earth, then there is no geological basis upon which we could conclude the world was billions or even millions of years old.

Uniformitarianism	Catastrophism
Geological Stability - The present is the key to the past. The processes we see happening today are sufficient to explain the Earth's appearance.	Geological Instability – large-scale disasters are primary in shaping the Earth and global climate. Exceptional circumstances have shaped much of the Earth's appearance.
Slow and gradual changes >> long ages (Small amount of water >> a lot of time)	Catastrophic changes >> short times (Large amount of water >> little time)
Implies Billions of Years	Implies Thousands of Years
*Uniformitarianism necessitates rejecting Noah's flood	*Biblical Catastrophism involves the flood of Noah

There is little debate over whether the bulk of the geological column is created under moving waters, because it is made of sedimentary rock (formed under water by definition). Additionally, it is also obvious that the geological column has been shaped by moving waters, as evidenced by many of Earth's features such as plains and canyons etc. Either way, water is primary in our understanding of the shaping of Earth's topography. Therefore, those who subscribe to Catastrophism generally use *Flood Geology* (a lot of water) to explain Earth's geological history. Flood Geology in fact used to be the prevailing scientifically accepted paradigm, but it was rejected in favour of an explanation that *excluded Noah's flood*. At the time, the Christian Worldview (being the basis of Science) was generally accepted by the culture. Obviously, not everybody was a Christian, but people generally accepted biblical history as history. However, the culture then began to become increasingly more secular (non-religious), and the Bible was largely rejected as a source document for historical truth. Many scientists then sought a scientific dealing with history, that did not involve accepting Noah's flood and the Bible as true history.

James Hutton is generally accepted as being the first to assert that Science needed to embrace millions of years of Earth's history. This was originally met with some opposition from the church community, which accepted his views as in conflict with the Bible. It was Charles Lyell who really was the father of Uniformitarianism. His famous book, *Principles of Geology* (1830), became exceedingly influential in convincing people that the present *is* the key to the past, and thus the idea of a global flood lost popularity in favour of a history of millions of years of slow and gradual processes. Lyell was intentional about opposing the biblical narrative because he wanted to *"free the sciences from Moses"*[108] - Moses was the writer of Genesis where the story of Noah's flood was told. However, Geology was still a fairly new endeavour at the time, and little was known about rocks and fossils. Since that time, Uniformitarianism has become the prevailing view in Geology, not because Flood Geology was refuted, but simply because it was no longer *popular*. Thus, all of historical geology that depicts the world as being millions of years old is based on an unproven assumption which was developed intentionally to be an anti-biblical narrative of history.

Therefore, when it comes to considering how old the Earth is, the real question is whether or not there has ever been a catastrophic event like Noah's flood or not. If there have *never* been such exceptional events in Earth's history, then Uniformitarianism is more accurate, and a time-scale of billions of years must be accepted. On the other hand, if a global catastrophe is the major cause of Earth's geological features, then the evidence previously seen to support a long geological time-scale is washed away in the deluge. Such a deluge is consistent with the flood of Noah as depicted in Genesis, and we can expect a time scale consistent with biblical history. This would suggest a geological column somewhere in the order of 4500 years old (the biblical date of the flood), with an ultimate age of the Earth of about 6000 years.[109]

---

108 Charles Lyell writing to George Poulett Scrope in June 14, 1830. See John Murray, *Life Letters and Journals* (1881) Vol. 1, p.268.
109 Noah's flood essentially cleared the slate and gave the planet a new beginning. Biblical Catastrophism sees the geological layers as being laid down in the flood, which also destroyed plant and animal life. Thus, most of what we observe (except perhaps the atmosphere and pre-Cambrian rocks) then dates to the time after the flood.

Uniformitarianism is a philosophical assumption which is typically not even supported by argument (in Geology textbooks), but merely assumed and presented as fact. One reason why Uniformitarianism is so central to Geology today is that it makes it possible to extrapolate physical process backwards through time to catch a glimpse of the past, as well as make inferences about the future (e.g. global warming[110]). It has also become the more desirable interpretation in the scientific establishment because it supports the evolutionary paradigm. That is, it would be impossible to formulate evolutionary theory based on a recent creation and a global flood.[111] However, the evidence that is usually given to support the assumption of Uniformitarianism is as follows:

(1) Current Processes Explain Everything – The key argument for Uniformitarianism is that the processes we see happening today *are sufficient* to explain *all the geological formations* that we see. (Obviously, if this was not the case then we would have to invoke exceptional forces not operating today, i.e. catastrophe.) If this is true, then we are forced to concede that these slow and gradual processes must have been in operation for many millions or billions or years to account for everything we see today. It is argued, for example, that the currently observed processes of sea floor spreading and continental subduction, geological uplift and subsidence, water erosion and sedimentary layering, when extrapolated over an enormous period of time account for all of Earth's current topography.

☒ It is merely an assumption that the current rate of these processes has been constant throughout time. Instead, it is more likely that because *"the energy of the Earth is less now than it was in the past, we cannot reasonably expect all geologic processes to have maintained their intensity during several billions of years"*[112], (even evolutionary geologists acknowledge this). For example, key processes such as volcanism, mountain uplift, and continental erosion may well have been higher in the past, and this cannot be ruled out simply because these processes are subdued in the present age. We simply don't *know* that current speeds for these processes have been constant, and therefore we cannot assume this is to be unconditionally true (we cannot assume Uniformitarianism).

---

110 Many of the arguments for global warming involve a look back into history, and this, it is argued, demonstrates the huge difference in climate since man's input. Most often, people are presented with millions of years of relative climate stability up till the present time (a Uniformitarian interpretation).

111 There would not be enough time for species to evolve. And the fossil record, interpreted under Catastrophism, would clearly Support Creationism rather than Evolution.

112 Quoted from a Geology textbook which supports Evolution. This is an example of when scientists contradict their own conclusions because of their preconceived philosophy. Edward A. Hay & A. Lee McAlester, *Physical Geology – Principles and Perspectives*, 2nd ed, 1984, pg 16.

- This argument seems to ignore slow and gradual processes which oppose the millions of years scenario. For example, processes such as the erosion of Niagara Falls (which has been measured) suggest a geological age of only thousands of years.[113] If we are going to assume that the present processes have remained at their current speeds, then we should arrive at an estimate of the world being less than 10,000 years old, because 90% of the current processes we observe suggest this! (See section below: 'Re-dating the World').

- The biggest problem with the Uniformitarian assumption is that there are many examples of Geological features that are *not in formation today*, and for which there are no credible scenarios, consistent with Uniformitarianism, accounting for their formation. Slow and gradual processes simply do not account for many geological features we see today. For example, the formation of billions of fossils, the stratification of worldwide rock formations, formations of bent and warped strata, polystrate fossils, fossil graveyards, the formation of large-scale seams of coal, gas and oil, the formation of large plains of strata, the transportation of rocks across country, the formation of water gaps, submarine guyots, and pediments etc, *are all things which we do not see happening today*. There are many thousands of examples of these, but there are no mechanisms currently in operation that could account for them. For example, unless water can flow uphill, it is impossible for 'water gaps' to form by any process currently in operation.[114]

- The kind of rocks we see forming today are not the same as the ones in the geological record. For example, we do not see the formation of diamonds, opals, vast deposits of fossils, and warped and twisted strata. If we do not see such things happening today, then it is unrealistic to conclude that the present is the key to the past, and we are forced to invoke exceptional circumstances to account for the observations. People could argue that we don't see this happening *because* it takes millions of years. Yet processes such as the formation of oil, stalactites, fossils, opals, and stratified rock have all been shown to occur rapidly under the right conditions.[115] (Conditions a global flood would provide!).

- One of the only examples of sedimentary rock layers forming naturally was seen in the aftermath of the Mt. St. Helens eruption in 1985, when moving waters and mud formed hundreds of alternating rock layers. These layers were formed in just three hours, and are very similar to

---

113 Based on observed erosion rates, the gorge created by the Niagara Falls is estimated to be in the order of 4400 years old. See Larry Pierce *Niagara Falls and the Bible* www.creation.com/niagara-falls-and-the-bible#f14

114 'Water gaps' are examples of where water has cut through the tops of obstacles such as mountains, rather than going around the base.

115 See Geology Questions and Answers at Creation.com: www.creation.com/geology-questions-and-answers.

common Geological features. The formation was exposed when a natural dam formed in the eruption was breached, and the water carved out a canyon. Observation of the formation of this feature has confirmed that stratification and the formation of canyons can happen rapidly and catastrophically. There is also good scientific evidence for this being the case in other areas, such as the Grand Canyon.[116]

"[M]odern geologists handle the many exceptions to uniformitarianism under the heading of "actualism," which could be defined as "uniformitarianism except when the evidence demands otherwise." They have to admit that the rock record often demands processes that are no longer observed and rates of processes that are much greater than present rates."[117]

"The present widely accepted system of Uniformitarianism in historical geology, with its evolutionary basis and bias, has been shown to be utterly inadequate to explain most of the important geological phenomena. Present rates and processes simply cannot account for the great bulk of the geological data. Some form of Catastrophism is clearly indicated by the vast evidences of volcanism, diastrophism, glaciation, coal and oil and mineral deposits, fossilization, vast beds of sediments, and most of the other dominant features of the Earth's crust."[118]

(2) Superposition – The assumption of superposition basically involves interpreting geological formations to be laid one on top of the other in decreasing age, so that the uppermost strata are millions of years younger than the ones on the bottom (the bottom was there first). Now, if the lower strata are much older, then we would expect to find a change in fossils as we get lower, ultimately finding organisms which died millions of years ago towards the bottom. Uniformitarian Geologists argue that the sequence of fossils that we find does confirm stratigraphic superposition over millions of years, as evidenced by the fact that the lowest strata contains fossilized species from early on in the *evolutionary sequence*. Therefore, what we see is that the rocks capture the progression from simple to complex organisms over time. Thus, because Evolution must have taken millions of years, the geological column has likewise taken millions of years to form.

☒ This reasoning is based on the *assumption of an evolutionary* scenario, and remember that the evolutionary interpretation of the fossil record is also based on the *assumption of geological time* and superposition![119] The fossils only support Evolution if the geological column was formed

---

116 See *Grand Canyon – Monument to Catastrophe*.
117 Mike Oard, *Flood by Design* (2008), pg 79.
118 John C. Whitcomb & Henry M. Morris, *The Genesis Flood* (1961) (36th printing 1992), p 439 . The last 400 pages preceding this quote gives the evidence and reasoning for their conclusions in this ground-breaking book.
119 See Chapter 2 (of this book)

over millions of years, and superposition only supports Uniformitarianism and millions of years if Evolution is true. Therefore, since this is a circular argument (it is self-referencing), it does not provide external evidence *for* uniformity and millions of years – it only assumes it. Obviously, if Evolution never happened then this argument would totally collapse (there would be no fossil basis for superposition).

☒ Geologists use 'indicator fossils' to identify certain strata and place those strata in the geological sequence. That is, the types of fossils which are found in the layers of rocks show where in the time-scale those rocks fit. e.g. Evolution sequence: fossil a/b >> rock layer a/b. Rock layers are organized into respective time-slots by '*indicator fossils*' which have an assumed evolutionary sequence/age. Therefore it is circular to argue that the sequence of rock layers is confirmed by Evolution, when that sequence itself has been organised according to Evolution! One evolutionary geologist put it like this: "*It cannot be denied that from a strictly philosophical standpoint geologists are here arguing in a circle. The succession of organisms has been determined by a study of their remains embedded in the rocks, and the relative ages of the rocks are determined by the remains of organisms that they contain.*"[120] If these rock layers were organised (in our textbooks) according to creationist assumptions[121] it would *not have the appearance of millions of years*. Thus, fossils do not support superposition – they fit in with it only if we assume Evolution and Uniformitarianism and sequence the fossils accordingly.

☒ Indicator fossils place certain rocks in their time-period according to the evolutionary scenario. This could be an illusion, considering that "*[I]t rather often happens that the fossils appear to be in reverse order from that demanded by the evolutionary history...*",[122] and also that some of the strata we find on the *surface of the Earth* is claimed to be hundreds of millions of years old. In other words, if supposition is true, and the oldest rocks are at the bottom, how is it that we find some of the oldest layers on the top? For example, in the Catlins of New Zealand there is an exposure said to be from the Jurassic period by the indicator fossils. However, how do we know that these fossils (and hence the rocks) are not much younger? This area is regarded as an example from this time that has been exposed due to the removal of the mountain of rock which was once on top (the more recent rock layers). However, if we were to make a different assumption – say that dinosaurs (fossils from that time) lived in the recent past - then finding Jurassic fossils

---

120 R.H. Rastall, *Geology* – article in *Encyclopedia Brittannica*, 1956, pg 168, Vol 10. As quoted in *The Genesis Flood*, pg 135.
121 Hydrodynamic sorting processes occurring under moving waters would usually result in the smaller and heavier organisms being on the bottom. Fish would naturally be on the bottom because the flood began with rising waters, leaving the most mobile animals last in the engulfing sediment.
122 Whitcomb and Morris, *The Genesis Flood* (1961), pg 135.

on the surface does not suggest Evolution or millions of years. The identification of those rock layers is dependent on the *assumption* of Evolution, and the evolutionary sequence is dependent on the assumption of superposition over millions of years. It is therefore again self-referencing to use one as evidence for another.

⊠ Whilst it may be true that the geological column shows small to large (assuming that the layers reliably show this trend), it is a misnomer to claim that this is 'simple to complex'. Some of the smallest creatures from the lowest rocks show amazing complexity. For example, the trilobite, commonly thought to be one of the earliest crustations, shows amazing complexity in its eye structures. Species that utilize such amazingly sophisticated technology for their basic functions would have to be considered 'highly evolved' by evolutionists, but yet we find these 'too early on' in the geological column to allow enough time for this to be plausible. Highly sophisticated organisms, which should be late on in Evolution, are inexplicably found in the earliest geological strata. Thus, even given the Uniformitarian interpretation of Geology, the rocks do not support simple to complex, and this is a problem evolutionists rarely communicate.

⊠ There are also known to be many anomalies in the geological column which raise serious concerns about its authenticity. For example, there are frequently whole geological eras missing from the sequence, or the rocks can even appear in a sequence which opposes the evolutionary paradigm! *"[W]hat should be said of the many examples of entire formations being out of place in the standard geologic time-table? In every mountainous region on every continent, there seem to be numerous examples of supposedly "old" strata superimposed on top of "young" strata."*[123]

⊠ The entire geological column has never been observed anywhere on Earth. It is well-known to be a composite image formulated from many observations in localised regions. However, even the most complete geological representations around the world comprise about only 10% of what should theoretically be present (according to evolutionary geologists), and these locations represent less than 1% of the land areas around the world. *"One unanswerable argument for the hypothetical character of the column is that nowhere in the world does the complete column exist. The majority of the geological periods are missing in the field."*[124] Without the assumption of Evolution and millions of years, there is simply no reason to suspect that the geological column, as depicted in evolutionary/geological resources, even exists.

---

123 Whitcomb and Morris, *The Genesis Flood* (1961), pg 180.
124 John Woodmorappe, *The Geologic Column: Does It Exist?* www.trueorigin.org/geocolumn.asp. First published in *Creation Ex Nihilo Technical Journal* 13(2):77–82, 1999.

(3) Radiometric Dating – The overall geological column is not found complete anywhere on Earth, but is pieced together by correlating the relative rock layers from different localities where only partial exposures are found. So for instance, when we find A-B-C then C-D-E we can formulate A-B-C-D-E. It is contested however, that radiometric dating techniques provide a numerical age for rocks, which transforms the geological record from merely *relative* rock layers into a *time sequence*. Thus, the 'absolute' ages, which are consistent with the evolutionary scenario, provide an independent means of determining the sequence of rocks other than the indicator fossils alone.

Sedimentary rock covers about 75% of the surface area of the Earth, and this is essentially the geological column we are dealing with. The ages of Sedimentary rocks are not determined radiometrically (and neither are the fossils found within them), but rather it is the portions of *Igneous* rocks and volcanic ash sandwiched between these layers which are dated. The isotope ratios of various elements within these rocks are accurately determined, and from these, an age is *inferred* based on assumptions about its history. The way radioactive decay works is that the original ratio of specific elements or isotopes in a substance changes over time. Therefore, given a good guess of the original ratio, the measured final ratio, and the known speed by which they transform one into another (the half-life), it is possible to determine the age of a rock sampled.

☒ Assumptions – All radiometric dating techniques are based on *unproven* assumptions about the past climate. For example, one has to guess the original amount or ratio of the particular isotopes, and this guess is based on a Uniformitarian assumption. That is, it is assumed that the key is the present to the past, and original ratios are based on what we *find today in our present climate* (assuming no climate changes). If we could not do this, then scientists would be 'flying blind' when it comes to comparing past and present ratios (a constant decay rate and absence of contamination must also be assumed). The other major assumption is the nature of the formation of sedimentary strata, which is not based on catastrophe, but on slow and gradual processes. For example, the volcanic ash found between those layers is caused by atmospheric volcanic events.[125] Of course, if this assumption is wrong, then the 'dates' obtained from radiometric dating can hardly be used to support millions of years - since this must be assumed to infer the date in the first place! In other words, with a different assumption in place, say, Catastrophism, there would be nothing about these ratios to suggest millions of years. The changes we observe in isotope ratios do not translate into dates when we see the entire geological column as being laid down by water in one event, because rather than being

---

125 If these ash layers were layered by underwater from underwater volcanic events (as predicted by biblical Catastrophism), then this would affect the calculations. For example, the amount of argon in the rock (used in K/Ar dating) would be different underwater.

indicative of the climate at that time, the composition of strata would be due to *hydrodynamic* sorting processes which occurred under moving waters. Radiometric dating supports millions of years only if we work off a strictly Uniformitarian (millions of years) framework in the first place, and hence is not evidence for that assumption (and the millions of years scenario which is assumed).

☒ Consider what would be the case if the pre-flood climate was quite different from our present, and Noah's flood caused a dramatic climate change.[126] The 'present is the key to the past' assumptions would lead us astray, for we could not arrive at correct assumptions by extrapolating back from today. For example, there is now mounting evidence that many decay rates could have been much higher in the past. Much work has now been done with 'radiohalos', the nature of which imply there must have been far higher nuclear decay rates in the past.[127] Other evidence comes from zircon crystals, which have been found to contain millions of years' worth of nuclear decay by uniformitarian assumptions, but at the same time have retained large amounts of helium (a by-product of nuclear decay). The problem for the old Earth paradigm is that it is well established that helium cannot be retained in those same crystals for that length of time (it escapes rapidly). Interestingly, the time that the helium appears to have been leaking from the crystals was measured to be 6000 ± 2000 yrs.[128] Therefore, there is mounting evidence that shows we cannot assume there has *always* been a constant decay rate, and this assumption is necessary for the radiometric method to work.

☒ The Paradigm of Evolution – Radioisotope data is well-known to *not be an exact science*, and for this reason it is common practice to discard any 'dates' which do not fit into the scientist's preconceived ideas of how old something is (they assume that the sample could have been contaminated). This is the case because there is no way of knowing for certain how old something is (because no one was there to observe it), and therefore there is no *verification* for any given age. Thus, what is accepted *by the majority* is the only thing close to verification as the approximate age. The only way, then, to really know if the method works is to check things of known age from *observed events*. In such cases, we find that radiometric dating methods *frequently give erroneous results*. For example: Mt Ngauruhoe in New Zealand erupted in 1949 and 1954, but those new rocks were 'dated' as being between 270,000 and 3,500,000 years old. A similar result was found for the rocks formed in the 1985 Mt St Helens eruption, which were dated by

---

126 There is evidence for this in the fossil record. For example, there were many examples of gigantism in plants and animals, and indications that things also lived longer in the pre-flood (fossilized) world.

127 Dr Russel Humphreys, *Evidence for a Young World*. See also Dr Don DeYoung, *Thousands Not Billions*, 2005.

128 Dr Don DeYoung, *Thousands Not Billions*, 2005.

the K-Ar method to be about 350,000 years old.[129] Samples of volcanic rock from the Grand Canyon thought to be 1 billion years more recent than the underlying basalt gave 'dates' of 270,000 years *older* – which is impossible because it suggests the lava flow is older than the rock it flowed over![130] Therefore, radiometric dating methods tend to rely on the paradigm of Evolution to give approximate ages, and therefore provide an internal check. Geologists are flying blind, unable to evaluate the validity of historical dates except by assumptions based on Evolution. When we can know the actual date of something by observation, we often find that the method is flawed.

☒ Different radiometric dating techniques are also known to give discordant results. Again, the dates in textbooks are those dates which are selected to agree with one another (because there is no independent means of verification), but the raw data can tell quite a different story. For example, basaltic rocks from the Grand Canyon gave different dates depending on the four methods used: K/Ar methods gave 10,000-117 M yrs, Rb/Sr methods gave 1,270 M -1390 M yrs, Rb/Sr isocron gave 1,340 M yrs, and Pb/Pb isocron gave 2,600 M yrs.[131] Uraninite crystals from Australia dated at 841 ± 140 M yrs by Pb/Pb isocron, 1550-1650 M yrs and 0-275 M yrs by other isotope ratios. The 'RATE' team of scientists report in the book *Thousands not Billions* at least ten examples of discordant results from independently tested samples, and overall, *"The results show clearly that discordance exists among the various radioisotope dating methods."*[132] Dates can hardly be relied upon when the raw data is self-contradictory.

☒ ¹⁴C dating is based on assumptions about a stable climate which are *known to be untrue*. ¹⁴C is produced by cosmic rays entering the atmosphere, and these are known to vary according to the intensity of the Sun's light, and the passage of the Earth through magnetic clouds in the solar system. Another factor that affects cosmic radiation is the strength of the Earth's magnetic field, which is well-known to be weaker now than in the past (this would make things look older according to the ¹⁴C dating method). The balance of ¹⁴C /¹²C in the atmosphere is also affected by temperature changes and increased volcanism, both of which have not been constant in the past. Additionally, if we did work on the assumption of a global flood, then the whole ¹⁴C dating system would need to be re-calibrated because of the burial of enormous amounts of ¹²C in the flood, and with the continued production of ¹⁴C, the balance of carbon would be very different from today, causing many

129 See Jonathan Sarfati, *Refuting Evolution* (1999), pg 110.
130 S.A. Austin, *Grand Canyon: Monument to Catastrophe*, Institute for Creation Research, Santee, California, pg 120-131.
131 S.A. Austin, *Grand Canyon: Monument to Catastrophe*, Institute for Creation Research, Santee, California, pg 120-131.
132 Dr Don DeYoung, *Thousands not Billions*, pg 110.

things to be estimated as much older by this method.[133] Overall, the ^{14}C dating method is dependent on a number of variables, and given that these factors are known to have been different in the past, ^{14}C dating can be *wildly inaccurate* and aberrant the further back in time we go.

Given that it is impossible to go back in time and observe what has happened in the geological past, scientists are forced to work off one assumption or another. Uniformitarianism is popular (being secular) and useful because makes it possible to extrapolate back in history based on current measurements. However, we have seen that the assumption of Uniformitarianism is just that, an *assumption*. The evidence used to support this philosophy is dependent on the philosophy being true in the first place, and therefore it is inadmissible as evidence for it. In other words, despite it's popularity, there really is no evidence supporting the assumption of Uniformitarianism, and in many cases, the evidence defies the assumption.

"When you examine the history of the rejection of the Genesis flood, you will discover that the rejection was not because of scientific reasons. A global deluge was never proven wrong. ... [T]he Genesis flood was outright rejected, not because of factual data or even superior reasoning, but because the intellectual elite abandoned the biblical accounts."[134]

"Uniformitarianism, in other words, has simply been assumed, not proved. Catastrophism has simply been denied, not refuted. But as a matter of fact it is not even true that uniformity is a possible explanation for most of the Earth's geological formations, as any candid examination of the facts ought to reveal."[135]

Evidence for a Global Flood

In short, it would take a large scale aqueous catastrophe (of biblical proportions) to explain the formation of large-scale geological features all over the world:

(1) The Geological Column is Laid Down by Water – Most of the Earth's crust is formed from alternating layers of sedimentary rock, which, by definition, were formed under water. Therefore, it is not contested that every landmass was at one time under an ocean of moving waters, and this also includes all of the highest mountain peaks such as The Andes and Himalayas (where we can find fish fossils). It is usually argued that landmasses have experienced *ongoing* uplift from, and subsidence into, the sea over Earth's massive geological history. However, given that we find matching geological layers all around the World, it is more consistent

---

133 See *The Answers Book,* pg 67.
134 Mike Oard, *Flood by Design* (2008), pg 31.
135 Whitcomb and Morris, *The Genesis Flood* (1961), pg 137.

to argue that these layers were formed *at the same time* in a single event (that all the Earth was under the same moving waters). Many of these geological layers are huge flat surfaces, such as the Coconino Sandstone located on the Colorado plateau which covers 500,000 square kilometers. *"It would take a large-scale, powerful current to winnow out the sand and spread it so evenly over such a large area."*[136] Such layers are compatible with others like it around the world, indicating that the most likely origin was a continuous uninterrupted event. Also supporting this, many geological layers show evidence of being laid down *at the same time* (despite being labelled millions of years apart by evolutionary geologists). For example, there are many examples where *multiple layers* of stratified rock are *tightly bent* without showing signs of fracturing, and this indicates that *all the rock layers* involved must have still been soft at the time of being bent. Obviously, if the rocks on the bottom were much older than the ones on the top, we would expect there to be much fracturing in at least those lower layers. We also find there is insufficient evidence for weathering / erosion between many rock layers, or any soil layers, or disturbance of roots and worms – which we would definitely expect to see if these really were at the surface for many years. There are also many examples of 'polystrate' fossils, such as tree trunks and large animals, which impossibly span 'millions of years' of rock layers. There is much evidence like this which indicates the geological column is more likely to have been formed in a single aqueous event. The major geological column we are familiar with is not currently forming anywhere on Earth. What we see is far more consistent with a massive flooding event.

(2) The Geological Column is Packed with Fossils – Usually when plants and animals die, they are broken down by bacteria in a matter of weeks[137], but fossils are formed only under very rare and specific geological conditions. Fossilization occurs when a plant or animal is buried in a cement-like mixture of grain and minerals which harden into rock, and during this process, the specimen is infused with minerals which eventually turn it into stone. For this specimen to be preserved in this manner (rather than breaking down), exceptional circumstances must prevail, and of course, if it is a dead animal, it must also avoid being eaten by scavengers. Fossilization conditions under water involve significant pressure, raised temperatures, and the exclusion of oxygen and bacteria. With such specific circumstances needed for fossilization, it is surprising that we find billions of fossils, *most of which are fish*, all over the world in such a well-preserved state – especially when we do not see many (if any) fossils being formed today.[138] For example, there were millions of bison killed in northern America in the last few centuries, but no fossilized

---

136 Mike Oard, *Flood by Design* (2008), pg 27.
137 Even shellfish, which comprise some 95% of all fossils, have organic matter holding their shells together, and when this degrades, the inorganic material falls apart.
138 Some fossils are forming naturally today. For example, when animals fall into tar pits they may fossilize. This is distinguished from those we find sandwiched in sedimentary rock.

specimens *have ever been found*. We do not see fossilization happening today, and certainly not on the sort of scale as the present geological column. Many fossils are very well-preserved, and show evidence that they were buried and petrified *extremely quickly*. For example, fossils of animals giving birth, fossils of one animal eating another animal have been found, and soft materials such as the fins of fish and whole jellyfish are also evident (which lose their fine structure very quickly). There are also many examples of 'fossil graveyards', where people find large quantities of fossils piled together in heaps.

The exceptional circumstances required for such processes are unlikely to be met if we consider only the processes happening today. However, a large-scale aqueous catastrophe where those stratified layers were being formed under large volumes of moving water provides just the sort of environment where massive amounts of fossilization can occur very rapidly. The fossilization of man-made objects such as fencing wire, hats, teddy bears, and bags of flour etc are well documented to have occurred in short time frames (months or years). Likewise, stalagmites and stalactites can also form very rapidly under similar conditions (cooling or drying mineral-rich solutions) as fossilization. People used to think that it would take 1000 years for one inch of these to form, until they began finding examples such as the 50inch stalactites under the Lincoln memorial in Washington (erected 1952). A large-scale flood would indeed provide moving waters that stratify rock, bury biota, and compress them in mincral rich solution. 'Fossil graveyards' could be easily formed in areas where animals are sloshed into particular regions as the topography and currents lead. As those waters cool, we could expect to see mineralization of plants (coal) and animals (fossils), and precipitation of rock (stalagmites and stalactites) on a massive scale. A massive catastrophic event must be invoked to account for the massive fossilization we see in the fossil record (which is not being added to today.) Large-scale fossilization is not happening today anywhere on Earth.

(3) The Geological Column is Uniquely Compressed to Form Fossil Fuels - Coal, oil and gas are made from the thermal de-polymerization (and mineralization in the case of coal) of organic matter with heat and pressure in an oxygen-free environment. The formation of the massive quantities of fossil fuels we find all around the world requires the rapid burial of vast amounts (whole forests) of material under water with high temperatures and pressures, followed by encapsulation within thick (hundreds of meters) layers of rock. What we find in the field is huge seams of coal and thick pockets of oil and gas. We do not see such events happening today by current processes, and it is completely unrealistic to suggest that these huge reservoirs could be the result of gradual processes. Rather it is more likely that massive Catastrophic events provided very exceptional circumstances. It is known that the formation of these fossil fuels does not need to take millions of years. Oil for example, can be made in the lab in 15 minutes, and coal can be made in 4 – 36 weeks (at 150 C). Additionally, oil can be found underground at pressures of up to 20,000 psi, which is far higher than what we would expect if it had been there for millions of years. We also do not see precious stones like diamonds and opals forming today, and these have also been shown to be formed in little time with the right conditions.[139] The vast reservoirs of fossil fuels we find is consistent with massive phenomena that stirred up and layered (under moving waters) unprecedented amounts of sediment and encapsulated fresh biota in hardening rock. Such processes are not in operation today anywhere on Earth.

(4) Catastrophic Processes Have Eroded the Continents – massive amounts of water erosion has occurred on the continents. Slow processes we see happening today do not account for the many planation surfaces observable, which can be up to 5000km in length. Huge volumes of water must have once moved in sheets across the continents to grind such areas flat, and transport cobbles and boulders hundreds of kilometers across-country from their origins. Some of these boulders have 'percussion marks' indicating that they were travelling underwater at speeds of up to 68mph, which is three times faster than the fastest flash floods currently being observed.[140] Continental sized sheets of water (a receding flood) is also the only sensible explanation for erosional remnants such as anticlines, tall escarpments, eroded domes, and submarine guyots - none of which we see forming today. Other erosional remnants such as Inselbergs (Ayers Rock) are also not forming today, and are a mystery in Uniformitarianism thinking. Again, massive sheets of water eroding the continents aptly explains the presence of such enigmas. The surrounding area has been planed flat by moving water leaving only the rocky outcrops, which are also planed flat by raised flood waters moving over the crest. The massive amount of eroded material is deposited in the oceans and settles at the continental margins, where water speed would have reduced, and this explains the presence of offshore continental shelves. Current processes cannot be used to account for the occurrence and prevalence of these features, and furthermore, current processes generally destroy such features! For example, erosional remnants such as Devil's Tower (Wyoming, US) are being rapidly eroded over time by current processes. Therefore, dramatically different large-scale events must be conceded as the explanation for such features.

(5) The Earth is Shaped by Receding Flood-waters – Many of Earth's geological features indicate that water-levels were high above the continents in the past, and moved quickly off into the ocean basins. This accounts for water gaps and flat plains etc. As water-levels lowered, the sheets of water flowing over continents changed into channelised currents

---

139 Andrew A. Snelling, *Creating opals: Opals in months—not millions of years!* www.creation.com/creating-opals
140 See Mike Oard, *Flood by Design* (2008), page 57.

weaving over the landscape and scouring out valleys and canyons in a short amount of time. When Uniformitarianism was first introduced, people then began to argue for the idea that current rivers were responsible for creating valleys, and therefore it must have taken millions of years to form them. Nowadays, it is recognised that the shape of many water-valleys is consistent with catastrophic flows of water in the past, that have created the pathway for the present river to flow. For example, evidence from catastrophic events, such as the Mt. St. Helen's disaster of 1982, have been shown to create canyons with vertically walled sides which are very similar to what we find in other parts of the Earth. Erosion over time shapes these water valleys into a v-shape, in contrast to vertically walled valleys – which give the impression of being very young. Another example is pediments, which is when we find planation surfaces within a valley. They can only be formed by massive amounts of water scouring out valleys and leaving flat plains at the bottom. Pediments are not forming anywhere on Earth today, because they require massive water flows. In fact, pediments are mostly being eroded by current rivers. Another common feature that suggests massive catastrophes, is 'water gaps'. A water gap is when a river flows through a mountain or obstruction (Grand Canyon is the most well-known example) rather than around. It is impossible for these types of canyons to have been formed by the river that is now in it, because the river would have had to flow uphill to begin with, or over a mountain rather than around it! Instead, we see that water gaps are evidence of a water level flowing above the height of such obstacles, and rapidly lowering and scouring out a pathway through them (as well as other waters going around). Thus, water gaps and the rivers that flow through them are best explained as an artefact of very high levels and flow speeds of water in a past event(s). Additionally, submarine canyons are another common feature of the ocean basin which are not being formed today - as far as we know. These can, however, be explained by massive and rapid continental erosion we would expect in a global flood situation, where material dragged out to sea carved through the freshly formed (soft) continental shelf formations (made from the sheet-flow phase of the flood).

(6) No Surface Structures Seem to Pre-date 4500BC – When we study the Earth, if it really was millions of years old, it is very strange that there seem to be no structures, civilizations, cave paintings, tree rings, or agricultures older than about the time that the biblical flood would have been. The scope and scale of the biblical flood suggests that it would indeed level all man-made structures and uproot trees burying all life under mountains of rock (which would fossilize and turn to coal and oil). The idea of the flood was for mankind to start again. The fact that this appears to be the case is strong evidence suggesting that something on the scale of the Genesis flood really did occur in the not too distant past.

Summary

Overall, there are many reasons why it is necessary to accept that there was a large-scale aqueous event (matching Noah's flood) that has shaped Earth's geography in the recent past. It is certainly more reasonable to accept this than the idea that current processes operating at their current rates are responsible for all of Earth's geological structures – when there is so much evidence to the contrary. If there was a year-long global flood that happened today, then scientists who survived would expect to see evidence matching what we already see on the Earth now. The assumption of Uniformitarianism cannot be sustained, and therefore the scenario of millions of years of geological time has no logical or factual basis.

Re-Dating the World

How is the World Dated?

The world simply cannot be dated by measurement because it is impossible to *measure* time that has already passed. Any discussion about the age of the Earth is based heavily on one's assumptions, because these estimations lack repeatability and falsifiability. That is, it cannot either be proven or refuted by experiment, because experiments are conducted in the present. Thus, the age of the Earth cannot be determined *empirically*, but can only be *inferred* based on the extrapolation of processes that we see happening today. The most reliable conclusion is the one which accounts for more of the evidence on hand. The trouble is, this inference is inevitably based on assumptions about the consistency of those processes, which in turn is based on one's preconceived view of Earth's history and creation. What someone believes to be true about how the world was formed directly influences the final conclusion from the data.

As we look at the world today, there are a number of processes occurring which have the potential to indicate the age of the Earth. What might be surprising to most people, is that somewhere in the order of 90% of these inferences are more consistent with the world being only a few thousand years old.[141] Again, it is because Evolution is the most popular view of origins that such information is kept a secret by the media and education systems. Below are some of those examples:

Inference Methods Implying Thousands of Years

Continental Erosion - At current rates of erosion by wind and rain it would take less than 10 million years for the continents to completely erode away to sea level. It is therefore estimated, according to geological time, that by now every continent should theoretically have been eroded

---

141 See Jonathan Sarfati, *Refuting Evolution* (1999), pg 112.

away about 100 times over. If the rock deposits containing dinosaurs really were millions of years old, the fossils should have been washed out to sea by now! Many of the world's mountains should also not exist. The average height of the landmasses of the world is about 623m, whereas estimated erosion over 2.5 billion years amounts to about 150km.[142] These observations are far more consistent with the biblical record that indicates the continents were formed in one event 4500 years ago (the flood), and have been eroding ever since.

Sea Floor Mud - Mud is also building up on the sea floor surprisingly quickly (at least 19 billion tons per year, because of erosion). At current rates the amount of sediment present would take about only 12 million years, yet the ocean is claimed to be 3 billion years old according to evolutionary thinking. If the oceans really were this old, they would contain sediment dozens of kilometres thick. A better explanation is that most of the mud was deposited there after continental erosion from the Genesis flood about 4500 years ago.[143]

Sodium in The Sea - Rivers pump 450 million tons of sodium into the oceans each year, and only an estimated 27% of this ever gets back out again. Even if the ocean was originally completely fresh water, it would have taken only 62 million years to accumulate all the salt that is currently in there (at a very conservative estimate). Estimates for some other (salt) ions give a much more recent value. This is much younger that the evolutionary age of 3 billion years. There is also additional geological evidence for much higher river flows in the past, greatly reducing this 'age'. Moreover, there is a large amount of erosion all over the Earth that indicates that an enormous amount of material has been dumped into the oceans at some time in the past. This, then, substantially reduces the estimated maximum age for the oceans - perhaps down to thousands of years. In any case, the oceans simply cannot be hundreds of millions of years old.[144]

The Earth's Magnetic Field - The Earth's magnetic field is decaying too quickly to have existed for millions of years. Scientists currently estimate that this field could not be more than 20,000 years old, because beyond this age the strength of the forces involved in generating the magnetic field would have been sufficient to melt the crust of the Earth. Hence life on Earth could not be older than this.

Biological Material - Some scientists have reported reviving bacterial from salt crystals found in rocks 'dated' at 425 million years old. Also

---

142 Tas Walker, *Eroding Ages* (2000), Creation 22(2):18-21, or @ creation.com.
143 D. Russel Humphreys, *Evidence for a Young Earth*, (booklet) 2005, Institute for Creation Research.
144 See also J. Sarfati, *Salty Seas: Evidence for a Young Earth*, Creation 21 (1):16-17, Dec 1998 - Feb 1999.

another find of bacteria from a bee's stomach was dated at 15-40 millions of years. DNA experts insist that DNA cannot exist in natural environments longer than 10,000 years, and this casts serious doubt on the validity of the dates given for these specimens.[145] Additionally, intact strands of DNA appear to have been recovered from fossilized Neanderthal bones, insects in amber, and even from dinosaur fossils.

Mutation Rates in People – A small strand of DNA is found in the mitochondia of human cells. This DNA is inherited directly from the mother, and therefore provides a good track-record of mutations. 'Mitochondrial Eve' is the nickname given to the theoretical first human female from which we are all related. In evolutionary thinking her age should be somewhere in the order of 1-10 million years. Measurements of the mutation rate of mitochondrial DNA recently forced researchers to revise the age of Mt-Eve from a theorized 200,000 years down to possibly as low as 6,000 years.

Helium – Helium is produced through radioactive decay and is found within Zircon crystals. There is enough helium in these rocks to suggest millions of years of radioactive decay at today's rates. However, helium is a tiny atom with a high diffusion rate, and in that length of time, the helium should have leaked out and no longer be present in those rocks. Uranium/Lead dating determines the 'age' of most zircon crystals at about 1.5 billion years. However, this appears seriously flawed because the amount of radiogenic helium present within the covalent lattice has indicated that the rocks could not be older than 5,680 +/- 2000 years. Since these zircons come from Precambrian rocks (the Earth's oldest rocks), that would also reduce the age of the Earth itself to about the same date.[146] This data also suggests there was an accelerated rate of nuclear decay at some time in the past. Additionally, if there really had been 3 billion years' worth of radioactive decay occurring (widely accepted), then we would expect about 1700 times the amount of helium in the atmosphere than there currently is, even when we take into account the amount of helium that escapes into space. Even if there was no helium originally, then the atmosphere could not be more than 2 million years old, by this reasoning.

Carbon 14 in Deep Strata - Coal, oil, and gas from around the world are now known to contain ^{14}C. This would be impossible if these deposits really were millions of years old, because the ^{14}C should have entirely decayed by this time. No source of coal has been found that completely lacks ^{14}C,[147] but yet, because of the relatively short half-life of 5400 years,

145 See *Bugs in Brine* Creation 24(4): 36-38, Sept 2002. *Isolation of 250 million-year-old halotolerant bacterium from a primary salt crystal, Nature* 407:897-900 (19 Oct 2000).
146 See *Radiometric Dating Breakthroughs,* Creation 26(2), 42-44, March-May 2004 and Jonathan Sarfati, *Refuting Compromise* (2004), pg 343.
147 *The Answers Book* (1999), pg 75.

^{14}C is virtually undetectable after a date of 50,000 years after the death of the plant in question. This finding effectively limits the age of all buried biota (all coal and oil etc) to less than 250,000 at most. Additionally, diamonds, which are supposedly billions of years old by evolutionary reckoning, have been tested for ^{14}Cl, and 'dates' of about 58,000 years have been inferred. Additionally, because these diamonds are found in what is regarded by (evolutionary) geologists as some of the oldest rocks on Earth, this ^{14}C find in diamonds necessitates an upper limit of the age of Earth's crust itself to be no more than this same age.[148]

Known History - According to evolutionary thinking, man appeared on the scene some 190,000 years ago, with a steady population of between 1-10 million people on Earth until about 10,000 years ago. If this was the case then we would expect about 8 billion bodies to have been buried over that time, yet only a few thousand have ever been found, and we would expect a larger human population today. Human history has been kept for only about 4000-5000 years, for example the Ebla tablets are some of the oldest history next to the Bible, and date at about 2300 BC. Agriculture, also, does not have any traceable record stretching back millions of years.[149]

Tree rings - The oldest known living trees have 4500-5000 tree rings, and are estimated to only be in their teenage years. Whilst it is possible for a tree to develop more than one ring per year, it remains a mystery (to anyone assuming millions of years of plant life) why no trees can be found with more than this number of rings.

Ice Cores - At the north and south poles there is a lot of ice (up to 10,000 feet holes have been drilled.) It is commonly believed that it has taken >50,000 years to have built up this amount of ice. Alternating layers of milky and clear layers of ice are assumed to be built up annually (like tree rings). However, 'The Lost Squadron' found in 1990 (lost 1942 - 48 years) was covered in 263 feet of ice, showing an average of 5.5 feet per year. At this rate, it would take only about 1824 years for the amount of ice to build up to 10,000 feet, not including the amount of ice from colder times (i.e. Ice Ages). It appears as though alternating layers of ice are caused not by the seasons, as once thought, but by fluctuations in precipitation.[150]

Human Population – Current population growth is about 1.7% per year. Now, if we assume an evolutionary scenario where humans had been around for about 1 million years, and a very slow average population growth since that time of only 0.01%, we would have 10^{43} people alive today, and we would expect about 40 billion buried human bodies somewhere! A scenario of 0.5% population growth for the last 4500 years

148 See *Radiometric Dating Breakthroughs* Creation 26(2), 42-44, March-May 2004.
149 D. Russel Humphreys, *Evidence for a Young Earth*, (2005) Institute for Creation Research.
150 See Carl Weiland, *The Lost Squadron* @ creation.com.

(since the flood) would give us a population today of 6 billion – a far more likely scenario.[151]

Coral Reefs – Coral has been found to contain growth rings, much like we find on trees, due to seasonal freshwater run-off. From correlating records of freshwater run-off with these rings, researchers have developed a new method of calculating the ages of coral. Based on this method, Australia's Great Barrier Reef is estimated to be less than 3700 years old.[152]

There are therefore a number of physical processes which suggest a young world in the order of thousands of years old.

Biblical Scenarios

If there ever was a world-wide flood, this would effectively be the beginning of Earth's history, because the flood would wipe off the face of the Earth all antediluvian (pre-flood) structures, people, plants and animals. Forming in its wake would be a renewed civilisation and ecosystem, together with freshly-laid geological features (this was the purpose of the flood). Human history and the like would therefore not be traceable further back than this time under this scenario (except for the History that Noah brought off the Ark with him[153]). As people grew in numbers on the Earth in the early days after the flood, the history of the flood would have been passed down the generations and be common knowledge. We in fact find that there are at least 500 flood legends from all over the world, and that these stories become closer to the biblical account the closer their origin is to the Middle East (where Noah came off the Ark).[154]

The Bible details the events of Noah's flood in chapters 6-9 of Genesis, but does not give many geological details. The Bible does, however, detail how there were 40 days of rain as the "fountains of the deep" broke open, and that for a period of 150 days the flood-waters rose high enough to cover every mountain (that existed at the time[155]) to a depth of at least 20 feet (the floating clearance of the ark). These waters then receded over the next 221 days, leaving the Earth once again habitable. The biggest geological question posed by the flood is firstly where all the water came from to cause the flood, and then where did it go to leave the dry land. The answer

---

151 See Don Batten (2001), *Where are all the people?*, @ creation.com.

152 See Paula Western (2002), *Coral: Animal, Vegetable and Mineral*, @ creation.com.

153 It is thought that the earlier chapters of Genesis were compiled from clay tablets that Noah brought off the ark. The repeated phases "This is the account of…" suggest the title of a new clay tablet.

154 See Mike Oard, *Flood by Design* (Green Forest, AR: Master Books 2008) page 25, and reference within on flood legends: T. Lovett, *Noah's Ark: Thinking Outside the Box* (Green Forest, AR: Master Books 2008).

155 It is thought that 'mountains' that existed at the time were relatively low, given that large mountains we have today are the result of continental shelf collisions. e.g. The Himalayas formed from the collision of the Indian and Asia continents.

is: from the ocean(s) and back into the oceans. Along with Genesis, Psalm 104 mentions an aspect of the flood of Noah that generates a workable Geomorphology.[156] It talks about how the waters of the oceans receded off the continents as there was continental uplift coupled with deep ocean basin subsidence. There is in fact a great deal of evidence that this has been the case in Earth's history, and that this is indeed a powerful explanation to account for general global Geomorphology.

The Flooding Phase

Most creationists conjecture that the ultimate cause of the biblical flood was catastrophic plate tectonic movements. Such a scenario might look like this: the existing super-continent[157] where all animals and plant life existed, catastrophically split open and began to move apart. This would have caused oceanic water to rush into the hot exposed mantle and evaporate, which would have generated a period of extremely heavy and continuous rainfall (40 days). Perhaps this could have been caused by the Earth expanding slightly in size.[158] At this time, the continent(s) subsided into the oceans and began moving across the mantle towards their current resting locations. This swamped all life on the Earth and buried it amongst vast deposits of sediment, which was sorted into alternating layers of rock under the moving waters (tides would have contributed towards this layering). These sediments were further compressed under the weight of the rising oceans, and this water, together with the subterranean heat, provided the right circumstances to form the vast deposits of coal, oil and fossils, as well as many of the precious (Metamorphic) stones we find today. Rapid continental drift continued, opening the ocean basins, until continents found their current locations or collided with other continents causing massive uplift of mountain ridges. Many freshly formed rock layers become twisted and buckled by these strong tectonic forces leaving today's stunning geological formations.

The Retreating Stage

This continued continental uplift coupled with ocean basin subsidence begins to move the water off the continents in sheets at first, then later in

---

156 "It is amazing how much support there is for Psalm 104:8" - Mike Oard, *Flood by Design*, pg 45. In this book Mike Oard develops a flood geology that elegantly explains many of the Earth's most common geological features that uniformitarianism geologists have spent that last 200 years perplexed about.

157 The Bible seems to imply in Genesis 1:9 that the waters were gathered in one place. Therefore there was one super-continent. This is also consistent with scientific theories.

158 In the book *Thousands not Billions* they propose that there was a period of accelerated nuclear decay sometime in the past. If this was on a large enough scale, it would result in increased mantle temperature (radioactive decay is thought to heat the Earth's core). Since the Earth's core is thought to be mostly molten metal (Iron), which expands with increased temperature, an increase in nuclear decay could expand the Earth's circumference generating catastrophic tectonic stresses. This accelerated nuclear decay could perhaps be linked with a time concentration event owing to the Earth passing through the Event Horizon in a White Hole Cosmology (see chapter 4 of this book).

channelised flows into the oceans where it is now.[159] Huge sheets of water recede off the continents at a high speed in a generally westward direction (because of Earth's spin – the Coriolis Effect). This causes the formation of large plained areas covered with smoothed pebbles, and also transports cobbles and boulders hundreds of kilometres from their origin. Eroded material is deposited at the continental margins where the water flow ceased, resulting in the formation of offshore continental shelves we find today. As water levels lower, uplifting mountains cause many continental divides, and water flow in many cases reverses direction as it is blocked by mountains and heads towards the nearest ocean. Sometime in this phase, Noah's ark runs aground in the Ararat ranges. A huge amount of erosion continues and increases in some areas as the water levels lower, producing eroded domes, anticlines, tall escarpments, erosional remnants (such as Devils' Tower) and Isenburgs (Ayers Rock). Water now drains off the continents in channels which carve vertically walled valleys, creates water gaps (e.g. Grand Canyon), tears pathways for rivers to flow, and forms pediments at the base of some of those valleys. Channelised material dragged out to sea finally carves through the soft continental shelf forming canyons perpendicular to the shoreline. The sum of this huge amount of continental erosion progressively turns the sea salty over the one year period, as ionic rocks are pulverised and dissolved in the ocean. Finally, as the land begins to dry out, vegetation returns from seeds and scraps left over from the flood. The wet rocks underfoot begin to dry, and underground caves are formed from the draining water.

The Aftermath

Assuming that the flood was caused by massive tectonic movements, and that mountain ridges were caused by continents smashing together, we can surmise that the crust we now live on is the broken and stressed remains of that catastrophe. In other words, the Earth's crust is *not* what it would look like if Earth had cooled from a molten ball of lava, but is covered in a crust that has been warped and twisted by massive past movements. We can therefore expect that the Earth's crust is fractured and unstable in many areas, especially those near impact zones between continents. Earthquakes, tsunamis and other volcanic activity are interpreted within a biblical creationist framework as being part of the aftermath of the flood.

The flood is thought by many creationists to have heated the oceans significantly (exposure of underground heat), and this in turn would have produced an increased level of evaporation and then precipitation at higher latitudes. Ultimately the oceans have the largest effect on global climate stability, and small changes there result in large-scale climate

---

159 This could possibly be the reverse of heat-forced expanding of the Earth that caused the flood. That is, as the Earth's mantel cools and contracts its 'stretched skin' buckles and warps (uplift and subsidence) - like the skin of a raisin.

changes. Thus, many creationists suggest an 'ice-age' lasting for a few hundred years, may have followed the flood. Under this scenario, Earth has undergone massive climatic changes in the recent past. It has therefore not been stable for millions of years, but in fact could still be bouncing back from hot and cold phases, caused by the flood, and the mechanism which caused it.[160]

Furthermore, the Bible hints there was no rain initially, but that the soil was watered by another subterranean mechanism (Gen 2:6). After the flood, God "placed the rainbow in the sky" and promised that the seasons for seed-time and harvest would continue. This could perhaps be suggesting that rain was not present before this time, and that the environment was temperate all year round, allowing for a constant supply of food. In any case, it is also thought by many creationists that the antediluvian (pre-flood) world was more hospitable to life. There is evidence for this in the fossil record, which has many examples of giant animals and dinosaurs, which must have lived longer that what seems to be possible today.[161] Inhospitable environments (such as deserts), have unfortunately also meant that many species of animals have not survived the last 4500 years to the present age, and famines and droughts are a reality. Some creationists, then, view extremes in weather as an aftermath of the devastation of the Genesis flood.

Thus, in the wake of the Genesis flood follows natural disasters, extremes in weather, famines and extinctions. These often called 'natural evils' are therefore not part of the originally "very good" creation that God designed, but are sobering reminders that we live on top of a world that was judged and perished under water.

What Happened to the Dinosaurs?

The 'Dinosaur' is a particular classification of reptile currently only found as fossils. So far as we know, they lived in the past, but are no longer alive today. The historical story of the dinosaurs as popularised in the education system and media is that dinosaurs became extinct hundreds of millions of years ago. This has posed somewhat of a mystery for scientists, for it is not known why a whole classification of animals had not survived to the present time, when so many other species have. Several scenarios have been suggested to account for the disappearance of the dinosaurs, the most popular being mass extinction from a large meteor impact.

---

160 'Global Warming' or 'Climate Change' unashamedly assumes millions of years of climate stability. Thus, recent changes in climate contrast sharply with this, and are therefore assumed to be because of man's influence. However, the less than 1 degree global temperature increase over the past *50 years* is less significant when viewed in the light of a massive, and recent global catastrophe.

161 See David Catchpoole, *Insect inspiration solves giant bug mystery* www.creation.com/insect-inspiration-solves-giant-bug-mystery. It is necessary for dinosaurs to have lived to an old age for them to be as big as they were. Also, the Bible also records people having lived to much older ages pre-flood. See Carl Wieland, *Living for 900 years* www.creation.com/living-for-900-years.

However, we ought to be aware of the fact that people only began identifying and classifying species of animals a few hundred years ago (the birth of Biology), and therefore many creatures could have been living right up to just before that time and we wouldn't know about it today! Global knowledge is a fairly recent invention also, thanks to transportation and telecommunications advances in the last century. This means that only in recent times has it become possible to have sufficient knowledge to identify such species as being extinct. This has also led to a growing awareness in recent times of conservation concerns. Hence, up to fairly recently, no one would consider it a crime to wipe out a particular species in a particular region, because you would have no way of knowing they were the last on Earth.

It is possible, then, that many species thought to have died out millions of years ago have survived right up until fairly recently in Earth's history, and there appears to be sufficient evidence to suggest that this might also have been the case with Dinosaurs!

Dinosaurs in Recent History?

Here is some of the evidence consistent with Dinosaurs being around in recent history:[162]

☑ Historical Evidence – There are many legends and stories involving 'dragons' which could well be dinosaurs. For example: China is renowned for its dragon stories and artwork, *The Anglo-Saxon Chronicles* records people encountering dragons, Alexander the Great apparently discovered huge hissing reptiles being kept in caves in India, Saint George slew a dragon that lived in a cave, a 10th century Irishman apparently encountered an animal not unlike a *Stegosaurus,* a 16th century scientific book listed several live animals we would recognise as dinosaurs, American Indians appear to have had a long history of encountering *pterosaurs,* along with the ancient inhabitants of Mexico and South America.[163] Aboriginal stories and cave paintings depict dinosaurs.[164] Descriptions of creatures matching 'extinct dinosaurs' have even been reported recently.[165]

☑ Artistic evidence – A brass plaque from 1496 in the floor of a church has pictures of dinosaurs etched into it.[166]

---

162 See *The Answers Book* (1999) chapter 19, or *The Creation Answers Book* chapter 19 available at www.aufiles.creation.com/images/pdfs/cabook/chapter19.pdf, and also the Dinosaur section at www.creation.com/dinosaur-questions-and-answers
163 See Bill Johnson, *Thunderbirds* www.creation.com/thunderbirds.
164 Rebecca Driver, *Australia's Aborigines ... did they see dinosaurs?* www.creation.com/australias-aborigines-did-they-see-dinosaurs.
165 Rebecca Driver, *Sea monsters* www.creation.com/sea-monsters-more-than-a-legend, Robert Doolan, *Are dinosaurs alive today?* www.creation.com/are-dinosaurs-alive-today
166 See Philip Bell, *Bishop Bell's brass behemoths!* www.creation.com/bishop-bells-brass-

☑ Scientific evidence – *Unfossilized* bones of a T-Rex have been found that contains red blood cells and soft tissues, suggesting that it is not millions of years old.[167]

These findings give some credence to the likelihood of dinosaurs being a part of our recent history, rather than dying out millions of years ago, and also the idea that humans and dinosaurs have co-existed. In fact, the Bible also suggests this. In the book of Job, which is thought to be one of the oldest books of the Bible, two large creatures are described which bear no resemblance to those we can find living today.[168]

The Bible presents the history of animals this way: about 2500BC there was a global flood which killed every living thing that breathed the air through its nostrils (including dinosaurs). This global catastrophe drowned all animals (except possibly insects) and entombed them in sediment which was compressed under moving oceans of water. Plants were turned into seams of natural gas and oil, or tempered into coal if there was a high mineral content present. Animals were turned into fossils through compression, heat, mineral infusion, and preserved because oxygen and bacterial were excluded in the process. Fossilization is in fact a rare occurrence requiring some quite unusual parameters, but a global event such as Noah's flood provides ample opportunity for mass extinctions and 'fossil graveyards' (large piles of fossils gathered in a rock basin).

Two of every basic kind of animal (such as dogs, large cats, bears etc.) were preserved with Noah and his family on his floating zoo. The Bible suggests that up to this point in history, all animals were vegetarian and ate exclusively plants.[169] After being on the ark for approximately a year, Noah and the animals disembarked and began to spread across the Earth. At this time, small genetically isolated animal groups amplified certain traits through successive generations producing a variety within basic kinds (such as the different types of big cats and bears). However, not all animals have established themselves and survived to the present time, because the postdulivian (post-flood) world is thought by many to be a much more inhospitable place to live. For instance, there are now many

---

behemoths#r17.

167 Carl Wieland, *Dinosaur soft tissue and protein—even more confirmation!* www.creation.com/dinosaur-soft-tissue-and-proteineven-more-confirmation, and other references on the T-Rex find at Dinosaur question and answers menu www.creation.com/dinosaur-questions-and-answers.

168 Job 40:15-25,41. The Behemoth and the Leviathan bear no resemblance to animals living today, but are consistent with some of the fossilized dinosaurs that have been found. It appears that God here is citing these creatures as two of his greatest creations. It's a shame they are not around today.

169 It is thought by some creationists that the pre-flood (Antedulivian) world contained some plant species no longer available today, which provided better nutrition, meaning that it was unnecessary for any animal to eat meat. But this has become necessary in the light of the extinction of these plants.

deserts and areas that are bleak, frozen, or mountainous, which limit the life in such regions.

Since Noah took two of every kind of air-breathing creature on the ark, Dinosaurs would have been included. A young breeding pair of dinosaurs would not have been very large (the largest dinosaur eggs were no bigger than a football). Dinosaurs, like other reptiles, would have kept growing until they died, and would get very large only if they lived very long. Dinosaurs came off the ark with the other animals and attempted to re-establish themselves. Several things could have prevented them from being as prevalent as other species. Perhaps the larger reptiles did not adjust well to the scarcity of plant food, and were also perhaps too slow to catch enough meat. If there was an ice-age at this time, reptiles in particular would have been restricted from moving out of the most tropical regions. In any case, Dinosaurs perhaps did survive in small numbers, but have become a part of the large list of extinct creatures. Besides, large dinosaurs would have been a menace to people, especially considering dragon legends, and were presumably hunted down and even eaten in times of famine.

Understood within a biblical framework of history then, there really is no mystery surrounding the dinosaurs.

Summary & Conclusion

The idea of millions of years comes not from observations, but from extrapolating back into history assuming that the present processes have been in operation for billions of years. Such a scenario is merely an assumed history called Uniformitarianism, which was designed as a secular replacement of biblical history. When challenged, this scenario is not found to be effective at explaining much of the Earth's geological features. The evidence used to support this scenario, such as radiometric dating (which has shown to be inaccurate anyway), is also grounded in the assumed truth of the scenario it tries to prove! It is therefore inadmissible as evidence supporting it. What we can know is that some form of catastrophic aqueous event must be invoked to explain what we now see, and such a scenario is consistent with biblical history. Additionally, there are many processes we see happening today which are more consistent with a young Earth, and even limit its age to only a few thousand years. Taken together, whilst we cannot measure time we were not there to witness, the concept of millions of years is simply wishful thinking and is not based on an evaluation of actual evidence.

The assumption of Uniformitarianism opposes the Genesis flood, and is necessary to support the time-scale of millions of years. Millions of years of time is a necessary assumption for Evolution, and Evolution in turn is a necessary assumption for Atheism. Therefore, those who are opposed to

God and/or the Bible are careful to guard the assumption of Uniformitarianism and the millions of years scenario of Earth's history. We have seen, however, that this assumption is without *scientific* grounding. Historically, the Genesis flood was never refuted, it just fell out of fashion with popular geologists. Hence, the millions of years' time-scale is nothing more than a philosophical construct, and one which defies most of the estimations we can make about Earth's past. People are of course free to believe whatever they like, but it is dishonest to teach such a philosophy as fact, when it most certainly is not.

In contrast, the biblical scenario centrally involving a Genesis flood is a powerful tool to explain the geomorphological processes that have shaped Earth's geology. For example, in his book *Flood by Design*, Mike Oard elegantly explains the formation of many of Earth's most common Geological features that have perplexed Uniformitarian Geologists for last 200 years. The biblical scenario also explains the 'mystery' of what has happened to the dinosaurs, and how we can reconcile natural disasters with a creation made by a perfect God.

Chapter 6 – Biblical History

Naturalism and Evolution are often celebrated but rarely presented with any supporting evidence. On the rare occasion when it is given, it is virtually a trivial exercise which shows the inadequate and non-scientific nature of those arguments. However, the concept of Evolution has been so popularized that many people think it is true simply because of its prominence in the culture and education systems. As well as many Atheists, there are also a number of 'Bible believing' people who accept many aspects of Evolution, and hence would attempt to marry Christianity with Naturalism. In this chapter it will be shown that the Bible is incompatible with Naturalism, and hence the true history of the World begins with God's creative work over a six-day period about 6000 years ago – as the Bible describes. In addition, since the basic 'anatomy' of Christianity is truth itself, it will be shown how evolutionary thinking is a poison to having faith in God.

Interpretation of Genesis

The Bible describes the creation of the world in simple terms in the book of Genesis. Over a period of 6 days, Genesis describes a number of individual creative acts culminating in the creation of man on day six. This event is expanded in Genesis 2, where the creation of male and female is described in more detail. God rests on the seventh day, and later this becomes the basis of the 4th commandment of the Law - where God himself writes on stone tablets instructing the Israelites to observe the Sabbath day as a way of honouring him for creation: *"For in six days the LORD made the heavens and the Earth, the sea, and all that is in them, but he rested on the seventh day."* (NIV, Ex 20:11). Genesis begins with man living in perfect relationship with God until the serpent deceives Adam and Eve into eating forbidden fruit, resulting in sin entering the world, and death through sin. The origin of death and suffering etc is therefore caused by man's general rebellion from God, and thus the solution to the world's problems begins with returning to right relationship with God. Thus, the Genesis account, particularly in Genesis 1-3, forms the essential basis of Christian Theology, by defining the place of God in the world, the origin, condition and destiny of man, the nature of sin and the origin of death and suffering, the basis of moral law, and the basis for marriage etc.

Common Reinterpretations

There are a number of common views of Genesis that attempt to reconcile the biblical account of creation with the concept of millions of years of geological time and/or Evolution.[170] However, all of these reinterpretations

---

170 See *Creation Compromises* (frequently asked questions) @ creation.com , or *The Creation*

either contradict the Bible, or Science, or even both:

(1) Theistic Evolution - Theistic Evolution assumes the truth of Biological Evolution and uniformitarian Geology, but at the same time affirms that God was somehow involved in this process. In Theistic Evolution, Creation is viewed as something that God set up from the beginning as a self-sustaining, *evolving process* extending over millions of years. Sometimes Theistic Evolution involves the concept of God 'guiding' Evolution from 'behind the scenes'. Man evolved from apes, but at a particular point in history, God put the breath of life in man, and hence created a spiritual being inside a human body. Theistic Evolutionists therefore often affirm the concept of there being a soulless race of ape-like men before the first *real* people were created in God's image.

☒ The first, and most glaring, problem with Theistic Evolution is that it is a contradiction in terms. Since Evolution is the Naturalistic explanation of origins, and God is supernatural by definition, the concept of Supernatural-Naturalism is self-contradictory. A *metaphysical* God could not have been guiding something which by definition is an *unguided* physical process. Theistic evolutionists must at least subscribe to a God who intervenes in creation to avoid this contradiction, but of course this would no longer be Evolution in any sense of the word. If God does create anything, then many of the evolutionary explanations for origins become redundant, because Evolution is an attempt to explain origins *without* the intervention of God. If God was not involved, then Evolution could not be called 'Theistic', but if God was involved, then it could not be called Evolution. If someone does believe that God exists, then why appeal to Evolution to explain origins?

☒ Another problem for Theistic Evolution is that the generally accepted evolutionary scenario is at odds with the Genesis account, and therefore there is no honest way to marry the two. For example, Genesis records God as *creating* animals according to their kinds in the same day, and hence there was no time difference or descent with modification. In fact, many of the events of creation described in Genesis defy the evolutionary picture of history:

Answers Book, chapters 2 & 3 which is also available @ creation.com.

Day	Event	Evolution Scenario
1	Earth created	The Big Bang produces Space - Galaxies and Stars first, the Solar System forms around the Sun
2	Sea and Sky separated (Atmosphere and Space created)	Earth begins to cool from a molten ball of lava first. Atmosphere forms
3	Land formed out of the Sea Plants created	Land first, then Oceans form. Non-living chemicals turn into Self-Replicating Systems then into Plants
4	Sun, Moon and Stars created in the Sky	Plants evolve into Fish
5	Fish and Birds created	Fish evolve into Land Animals, and then into Dinosaurs
6	Land Animals created Man created in God's image	Land Animals evolve into Mammals, Dinosaurs into Birds Man evolves in the image of a Gorilla

According to Genesis, apes and humans were created on the same day, and hence there is no progression from one to another. Genesis also says that God formed man from the dust of the ground (Gen 2:7), and to dust he would return (Gen 3:19). But, if we are to accept Theistic Evolution doctrine, that we evolved from a gorilla, does this mean that when we die we go back to being a gorilla? The Bible affirms that Adam and Eve were real people God created especially in his image for relationship with himself and each other. Jesus words affirm this happened "at the beginning of creation" (Mk 10:6), and Luke calls Adam the "son of God" (Lk 3:38). Paul also affirms that Adam was the first man formed from the dust of the Earth (1 Co 15:45-47). Thus, the New Testament also affirms the Genesis story as the true origin of man.

The first key premise of Theistic Evolution is that Evolution actually occurred and is scientifically credible, hence people try to fit Genesis in with that. This has already been shown to not be the case at all. In each of the four key areas of origins science, this book has shown there to be a complete lack of factual support for Naturalism through Evolution. The second key premise is that the Bible is compatible with Evolution. This is also not the case. If Evolution occurred, then it would certainly contradict the Bible and would not be supported by the Christian God.[171] If the Bible is true, then Evolution never occurred. The attempt to combine these

---

171 Since what he has said in scripture contradicts Evolution. If the Bible really is from God, then God is hardly going to support something that goes against his word and is not true.

conflicting viewpoints only generates false definitions of both. However, if there was no Evolution, and no other God aside from the God of the Bible exists, it might be safe to combine these, and assert that a non-biblical God caused nothing to happen!

(2) Progressive Creation - Progressive Creation[172] is the idea that God created the world in successive stages over billions of years. It therefore usually involves the general acceptance of the concepts of Big Bang cosmology, Geological Uniformitarianism, and Biological Evolution in some form, with the exception that these events involved the intervention of God throughout time (rather than things naturally evolving). In this view, the creation of the cosmos happens over billions of years, and therefore the biblical account in Genesis is not taken as historical, and the 'days' are often interpreted as overlapping indefinite periods of time. In accordance with Uniformitarian philosophy, the Geological column is interpreted as spanning millions of years and representing many successive events (of sedimentation and fossilization). Noah's flood therefore is regarded as only affecting a localized area, and so is still minor in significance. The fossil record is accepted as revealing a progression of organisms over millions of years, except that each new species discovered in it (fully formed) represents another *creative intervention* of God.

☒ It was a very similar view of Earth's history that was the motivation of Darwin formulating 'Evolution' in the first place, because he saw this type of Creationism as lacking the power to explain the actual observations. Progressive Creation events are not consistent with the variety and location of species of animals, and if we accept a local flood (which people tended towards in Darwin's day), then, as Darwin saw it, Evolution over millions of years is a more logical alternative. We have already seen, however, that the biblical history involving an ark which preserved animals from a global flood has a more powerful ability to explain the facts we observe today. It is also a misrepresentation of the Bible to claim that the flood was only local. For example, what would be the point of having an ark in the first place if people could simply flee to surrounding areas to avoid the flood-waters? If the flood was only local, that would also mean God was a liar when he said "...never again will there be a flood to destroy the Earth." (Gen 9:11 (NIV), see also 8:21 & 9:15), because there have been many local floods in history! The Bible claims the waters covered all of the highest mountains of the time (Gen 7:19), and is always referred to throughout the Bible as a cataclysmic event (even in the New Testament). Once Catastrophism (biblical history) is accepted as the correct interpretation of the Geological column, then the evidence obviously favours a recent creation and the progression of animals in the fossil record is not related to descent with modification.

---

172 See Jonathan Sarfati, *Refuting Compromise* (2004), and Van Bebber and Taylor, *Creation and Time* (1994). These are books dedicated to refuting Progressive Creation.

☒ We have already seen in previous chapters that it is scientifically inaccurate to accept the Big Bang, Uniformitarian Geology and the sequence of Evolution. We have also seen above that the events of creation conflict with the events of the evolutionary sequence. In addition to these, it is simply inaccurate to say that the God of the Bible created then rested for long time periods before the next instalment in creation. Genesis (and Exodus) says that God created everything in 6 days and then rested on the seventh day, thus God is still resting from his creative work because he accomplished it all in that one week (Heb 4:3).

☒ Despite its popularity, Progressive Creation does not represent good science nor good biblical exegesis.[173] Moreover, there is a specific warning about this kind of thinking in 2 Peter 3:4-6. Here Peter asserts the world was created from water and out of water,[174] and that by these same waters the world that once existed was destroyed. He also warns that people who are against Christianity would scoff at Christians, using the argument that things go on as they always have (Uniformitarianism), and therefore since God has never judged the world before, we can expect he never will. This passage then correctly predicts the rise of Uniformitarian philosophy in order to deny the past judgement of the flood (what modern 'Science' attempts to explain) and future judgement of the world by Jesus. Therefore, biblically minded people should be especially cautious about these philosophies.

(3) The Day-Age Interpretation - Some people have attempted to harmonize the Bible with millions of years of time by suggesting that the 'days' of Genesis really refer to longer periods of time, perhaps indefinite periods of time. Scriptures such as 2 Pet 3:8 and Psalm 90:4 are often referred to because they indicate timelessness from God's perspective. Thus, it is argued, from God's perspective the six days of Genesis could mean virtually anything – just not six actual days!

☒ The most obvious problem with this view is that it does not matter whether God is outside of time or not, because he has communicated his truth to we who are inside time. Thus, the debate comes down to whether God is a liar or not. This is usually rebutted by challenging the common interpretation of the Hebrew word for day, '*yom*'. However, the meaning of the Hebrew has much the same meaning as the English word day. That is, it can be used in three senses: (1) It can be used in

---

173 Exegesis is when we read something and gain understanding *out of* it. It is about comprehending the meaning of the text that an author intended to convey. The opposite, eisegesis, is when one reads something with the intention of finding agreement with one's own opinion.

174 This seems to be describing the creation of Earth out of an amorphous water-based universe, and then the formation of dry land out of the oceans (Gen 1). It was this oceanic water that destroyed the world in Noah's flood.

the indefinite sense (e.g. "in the day of my distress" Gen 35:3, "in the day of my disaster" Ps 18:18, (2) It can be used for the 12 hour period of the daytime (e.g. "in the heat of the day" Gen 18:1, "The heat consumed me in the daytime and the cold at night" Gen 31:40), and (3) It can be used to mean a 24-hour period of day and night (e.g. "Commemorate this day, the day you came out of Egypt" Ex 13:3, "It shall be eaten on the day you sacrifice it or on the next day". Lev 19:6). The appropriate meaning of the word in each instance is determined by the context of the sentence. When *yom* is associated with a number it is always the definite sense of 24 hours, because it would be nonsensical to interpret it as an indefinite-definite time period. For example, the Law stated that a woman was unclean for *7 days* due to her monthly flow of blood (Lev 15:19), but no one tries to argue that this means millions of years! Jesus was circumcised on the 8th day (Lk 2:21) according to the Law of Moses (Lev 12:3), and this shows that the early church understood the meaning of the word day when associated with a number to be 24 hours. *Yom* is used with a number 410 times outside of Genesis and it always means an ordinary day.[175]

☒ At the closure of each 'day' in Genesis it says *"And there was evening, and there was morning - the [nth] day."* This again shows that God here is talking about a *cycle* of time, and this cements the idea that it was a 24-hour (not 12 for example) sequence. This method of associating evening and morning together is used a further 38 times outside of Genesis 1, where, again, it always means an ordinary day as we understand it to be today. If God had intended for us to believe that he created the world in 6 ordinary days, it is hard to imagine how he could have made it clearer.

☒ Adam and Eve were created on day 6 (and lived through day 7 obviously), and Adam was recorded to live to the age of 930 (Gen 5:5). Interpreting the days to mean millions of years would also mean that Adam lived millions of years!

☒ Finally, the largest problem for Day-ager's is not so much involved in the interpretation of Genesis, but what to do with Exodus (Ex 21:11, 31:17). "In six days" is clearly stated there in terms of six ordinary 24-hour cycles of time, and it forms the (only) basis for our seven day working week. This commandment would have no meaning if it referred to indefinite periods of time, and if it referred to longer times (such as a million years each), then are we expected to work for six million years and then rest for one? God is not in any way limited to our time, but as we can see from the forth commandment, God evidently created the world in six ordinary days as a model for us to follow.

---

175 *The Creation Answers Book*, pg 39.

(4) The Gap Theory - Proponents of the 'Gap Theory'[176] adhere to the 'days of Genesis' being literal 24-hour periods of time, but also believe in deep time (billions of years). The way that they reconcile the two is to insert a huge time-period between a supposed initial creation and the six days of Genesis – that is, between Genesis 1:1 and 1:2 (or sometimes between 1:2 & 1:3). The idea is that there was an original creation which was perfect until Satan rebelled and God judged the world with a flood ('Lucifer's flood'). 'Gappists' translate Genesis 1:2 as "the Earth *became* formless and void"[177], and then the world was recreated in the six days that follow as per the rest of Genesis.

☒ The reason why people propose the billions of years before the 'reconstructed' creation is to settle with the evidence from Geology which they feel suggests such a massive period of time. However, one of the key problems with the Gap Theory is that it cannot be reconciled with popular Geology, because Uniformitarianism is dependent on there *not* being a large scale flood to arrive at the millions of years in the first place! In an attempt to reconcile with the fossil record, Gappists have also put death and suffering before the fall of man (a heresy[178]). The Gap Theory also violates the clear teaching of the 10 commandments, where the whole Heavens and Earth were created in 'six days'. Thus, the Gap does not agree with the Bible, nor with Geology, and it contradicts both.

(5) The Framework Hypothesis - Proponents of the Literary Framework Hypothesis regard Genesis as a work of poetry that was never intended to map onto reality (to be taken literally), but rather is a fictional/figurative document. Genesis is therefore not regarded as a historical document describing creation at all, but rather it is there to describe *why* God created and not *how* – since they claim that the Bible is not a 'scientific textbook'. In the same way as other myths of the time, Genesis communicates 'truths' without having to actually *be* true. Sometimes associated with this Framework is the concept that Genesis 1 & 2 are different (contradictory) accounts of creation (rather than the traditional view of Gen 2 as an expansion of details of Day 6 in Gen 1), and thus, also cannot be taken as historical.

☒ The key argument that such proponents make is that one cannot take every part of the Bible literally, but we need to pay concern for the genre of the text. For this very reason however, we ought to pay close attention to the fact that Genesis is a *historical narrative*. We therefore should treat it as recording events in history in the same way as we would treat any other historical narrative. For instance, if we are

---

176 See Chapter 3 of *The Creation Answers Book* for a refutation of the Gap Theory.
177 Which is grammatically unsound, see *The Creation Answers Book,* pg 59.
178 This is explained later in this Chapter.

going to disregard the historicity of Genesis, then why not do the same with the details of the resurrection of Christ, or the creed in 1 Cor 15:3-8? Genesis claims to be historical, and this is certainly the affirmation of the remainder of the Old Testament and the New Testament. There is simply no reason to accept that it is anything other than factual, and indeed this was the consensus on Genesis historically until this view was conceived in 1924.[179]

- ☒ Arguments pertaining to the poetic nature of parts of Genesis (such as 1:27 and 2:23) do not destroy the historicity of those statements – since something that is well-written can also be true. No one would use such an argument for any New Testament passage. Genesis 1-11 (the creation and flood accounts) forms a seamless continuum of history with other events, such as the destruction of Sodom, which are generally regarded as historical. It is also written in the same Hebrew structure as with the rest of the Torah, which is not considered to be allegory but history. In particular, Hebrew has a particular grammatical form for recording history as distinguished by its usage of verbs.[180]

- ☒ It was common for ancient writings (especially Egypt where Moses was educated) to begin with an overview of the events in their chronological sequence, followed by an expansion of particular events in finer details.[181] In Jesus' teaching on marriage (Mt 19:4, Mk 10:6) we see that he regards Genesis 1 & 2 as complementary accounts depicting the same events – which is necessary for the concept of marriage to have any real meaning.

- ☒ There are a number of other historical documents which depict events strikingly similar to Genesis. The Epic of Gilgamesh, for example, is an account of a world-wide flood. Judging from embellishments and absurdities contained within these stories, Genesis appears to be the original source document/tradition from which subsequent legends have developed. Flood legends in fact occur in almost every ancient culture around the world.[182]

Whilst the Bible is not a 'scientific textbook', the real issue is whether it is *true* or not. If it is, then when it records events as happening in space and time, then those events happened in space and time regardless of whether it is poetic or not. If it is not true, then it doesn't matter either way because the Bible would not be worth believing as authoritative.

---

179 *The Creation Answers Book,* pg 45.
180 See Boyd, S.W., *15 Reasons to Take Genesis as History, Thousands Not Billions* (Chapter 10), and *The biblical Hebrew creation account: new numbers tell the story.* Impact 377, 4 pp. www.icr.org/pdf/imp/imp-377.pdf
181 See Don Batten, *Genesis contradictions?* www.creation.com/genesis-contradictions.
182 Jonathan Sarfati, *Noah's Flood and the Gilgamesh Epic,* www.creation.com/noahs-flood-and-the-gilgamesh-epic.

Oxford Hebrew scholar James Barr wrote:[183]

"... probably, so far as I know, there is no professor of Hebrew or Old Testament at any world-class university who does not believe that the writer(s) of Gen. 1–11 intended to convey to their readers the ideas that
(a) creation took place in a series of six days which were the same as the days of 24 hours we now experience
(b) the figures contained in the Genesis genealogies provided by simple addition a chronology from the beginning of the world up to later stages in the biblical story
(c) Noah's flood was understood to be world-wide and extinguish all human and animal life except for those in the ark."

In conclusion, it is obvious that the only honest way to interpret Genesis is as a historical document, because all other variations end up in self-contradictions (within the Bible). This is indeed how Jesus and all authors of the Bible regarded Genesis, and is consistent with the bulk of the scientific data we've seen in previous chapters.

Why All The Different Ideas?

With all of these different interpretations of the creation account, we can readily see that none of them are consistent with the rest of the Bible. Obviously then, one cannot arrive at these positions *from the Bible alone*, and hence, it seems that influences from outside the Bible are playing a significant role in the way that people interpret biblical history.[184] In this case, evidently the Bible is being reinterpreted in the light of what people (falsely) believe to be the actual evidence pertaining to the history of the world, as conceived of by atheistic scientists and philosophers.

Whilst people might believe in the naturalistic scenarios of history, it is clearly dishonest to say that the Bible is compatible with them. When reading any document, it is dishonest to try to fit that in with your own preconceived views when these conflict with what the author was originally trying to communicate. When one bends and distorts a document to agree with a viewpoint one has already decided on to hold, then one raises *that* viewpoint as authoritative *above* what one is reading. Doing so with the Bible is to raise the opinions of naturalistic scientists above the Bible and Jesus' teachings. For a Christian, this then would constitute an *act of idolatry*, because it is raising foreign (to the Bible) beliefs to be authoritative above God.

---

183 Barr, J., in a letter to David C.C. Watson, 23 April 1984. Quoted in *The Creation Answers Book,* pg 38.
184 There is also significant historical evidence to show that these views were not around before the rise in deep time claims from scientists. This is strong support for the fact that these views are not born out of the Bible, but from an adherence to some scientists' claims. For example see *Refuting Compromise* chapter 3 and *The Creation Answers Book,* chapter 2 .

Tolerance and Interpretation

The creation event is not something we can observe as happening today, and because of the controversy over how the world came about, many people are unsure of the Bible's place in the debate. Hence, in general, many people it seems are open to accepting various conflicting viewpoints over the creation story in Genesis. Thus, the Bible is often seen as being capable of saying a number of different things depending on your 'interpretation' of it, and we are often urged to be tolerant of these different opinions. Of course, if we ought to be tolerant of *any* opinion, then we ought to also be tolerant of the opinion that only the literal interpretation of Genesis is correct! Thus, the argument affirming tolerance of conflicting views is self-defeating. An idea is either true or false – depending on its correlation to the facts, but mutually exclusive ideas cannot both be true.

Whilst some parts of the Bible can be interpreted differently or non-literally, that is not true for historical narrative. Parts of the Bible are highly figurative and do not allow for a literal interpretation, but this is quite different from reinterpreting passages which claim to be a record of things that have happened in history. The only correct way of interpreting historical events is historically (literally). Where the Bible records something as happening in space and time, it either happened that way or the Bible is in error. Conflicting claims about a historical reality cannot both be true, and so the real question is whether the Bible claims to record the true history of the world or not.

The context is the simplest way of determining the correct interpretation of something, and it is obvious that the early chapters of Genesis form a continuous historical account leading all the way through to the exodus of Israel from Egypt (which everyone accepts as historical narrative). Alternatively, we can look at other parts of the Bible and see how they viewed Genesis. A straightforward reading of Genesis as true history is affirmed by other scripture such as:

➢ Many passages in the Old Testament assume that Genesis 1-11 is a record of historical events. Six-day Creationism is a non-negotiable commandment in Ex 20:11, and Ex 31:17, both of which are claims to events that happened in history. The Israelites were required to believe that God made everything in 6 days as part of the law, because it was the basis for the 6-day working week. Jesus affirmed the Sabbath day of rest, thereby endorsing this commandment (e.g. Mk 2:27). Hebrews also quotes from Genesis 2 referring to Sabbath rest (Heb 4:4).

➢ The NT contains more than 100 direct references or quotations to Genesis 1-11 from every one of the NT authors. Genesis is quoted or referred to in both the Old and New Testaments more than any other book in the whole Bible – and in each instance it is taken as a historical account of actual events.

> The Bible itself claims to be the total record of human history 'from the beginning' of creation, and we see Jesus particularly affirming this stance in Luke 11:50. The genealogies found in Chronicles (1 Chron 1) and Luke (Lk 3) show that these writers affirm that the Bible records the total history going all the way back to the first person created by God. Jude affirmed these genealogies by saying that Enoch was the "seventh from Adam" (Jude 1:14). The writer of Hebrews (see Heb 11) starts history from the creation using the Genesis record as an authoritative source depicting the faith many OT people starting from Abel, Enoch and Noah (from Gen 1-11).

> Jesus affirmed the teaching of Genesis as authoritative historical truth. For example, he referred to Abel as a real person (Lk 11:51), referred to Genesis 1 & 2 as the basis for marriage (Mt 19:4, Mk 10:6) (along with Paul - Eph 5:31). He affirmed that Genesis 1 & 2 was referring to the same creation event of male and female (Mt 19:4). He even said that if someone could not believe in Moses (who wrote Genesis), neither then would they believe in him (Jn 5:46).

> The Bible consistently affirms that the flood of Noah was a global catastrophe which wiped out all life on Earth, and thus the ark was needed in order to preserve human and animal life. Jesus (Mt 24:37-39, Lk 17:26-27) and Peter (1 Pet 3:20, 2 Pet 2:5) in particular affirmed the historicity and epic nature of the flood.

> Paul held that Adam (Rom 5:14, 1 Cor 15:22, 45) and Eve (1 Titus 2:13, 2 Cor 11:3) were real people, and used the order of their creation (Gen 2) as a basis for teachings.

> The creation by God's word as a historical reality is affirmed in Heb 11:3, 2 Pet 3:5, Jn 1:3, Col 1:16.

These examples demonstrate that the rest of scripture affirms a straightforward reading of Genesis as historical Narrative – which is what it claims to be. That is, a recent supernatural creation of the Universe and man in 6 days is what the Bible affirms as true. As we have already seen in earlier chapters, this too is scientifically defensible.

Jesus on Creation

A Christian, by definition, is someone who follows Christ, and in particular, who follows his teachings. Jesus claimed to be God, and thus he claimed to teach what was authoritatively true for all people. His claims were centrally supported by various evidences: a perfect life of loving others, specific signs and miracles, the fulfilment of many specific prophesies, and the bodily resurrection from the dead. A Christian sees these evidences as good reason to place their trust in Christ, and follow his truth. Christ's teachings form the content of the New Testament, and he also affirmed the Old Testament as authoritative truth inspired by God. Therefore a Christian, by definition, holds the Bible as truth above their own or anyone else's opinion. To not follow this would be to say that we were smarter than God!

When Jesus was asked about marriage he responded this way: *"But at the beginning of creation God `made them male and female.' `For this reason a man will leave his father and mother and be united to his wife, and the two will become one flesh.'"* (NIV, Mk 10:6, see also Mt 19:4). Jesus here upholds the creation of both male and female in the image of God as the reason for them coming together. They did not evolve and then find each other, but rather they were created at the beginning of creation by God *for each other* – otherwise marriage would have no deeper meaning. Now let us compare this picture with the evolutionary time-scale:

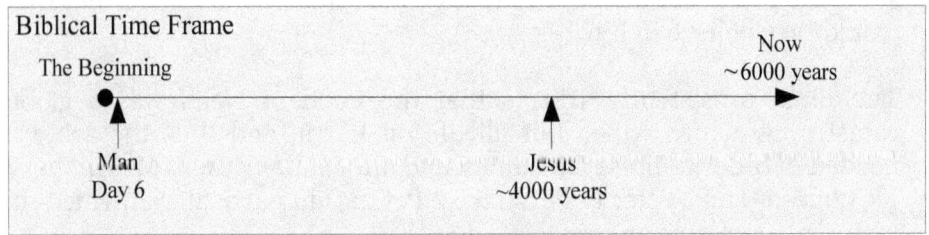

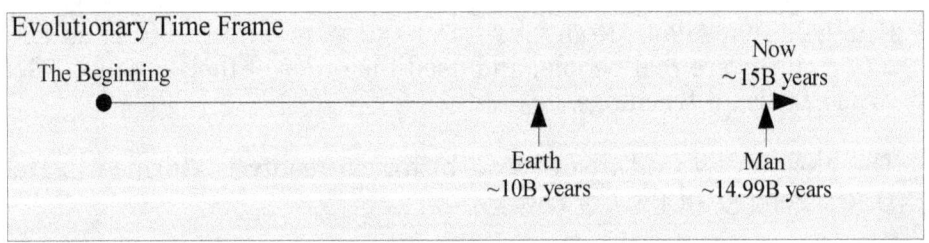

	Universe Began	First Man	Time After Creation	% Time Passed
Biblical Time-Frame	-4000 yr (before Jesus)	-4000 yr	6 days	0.0004%
Evolutionary Time-Frame	-15,000M yr	-5M yr	14,995M yr	97.97%

Under the Evolutionary time-frame, by the time man came on the scene most of the time had passed, and therefore man was not created at the beginning. Under the biblical 6-day creation event about 4000 years before Jesus, man was created on the sixth day - right there at the beginning. Therefore, only the idea of 6-day creation in a young Universe is consistent with Jesus' teachings here. He stated that man is as old as the world, and therefore there was no Evolution and no millions of years of time, according to his teaching. The concept of billions of years of time passing from the beginning until man is created/evolves is clearly in contradiction with Jesus' teaching, and as such should be rejected by people who claim to follow his teachings as authoritative.

The Anatomy of Faith

Heresy

A heresy is a belief that affirms a non-biblical distortion of Christianity, and usually this is treated seriously because it defeats key aspects of Christianity/salvation. For example, there are religious movements that still hold to the authority of the Bible, but yet because they affirm a particular person's (distorted) teaching as authoritative, they affirm a *distorted interpretation* of scripture. Some of these such teachings are heretical because they take portions of the Bible and force an interpretation that is not consistent with the rest of scripture. Thus, through questioning, it can be shown that the Bible is in fact a secondary document after that person's adherence to a particular teaching. In the end, heresy comes down to putting a person over God, because that person's view of scripture becomes the final authority on what someone will believe from scripture.

However, when it comes to Evolution, often it seems to go unnoticed that it is heretical to accept a distorted interpretation of scripture for the sake of staying true to 'Science' (opinion of popular scientists). For example, any one of the five reinterpretations listed earlier clearly conflicts with scripture in order to try to agree with what some see as 'scientific'. In this case, it is the science which is raised up as infallible, and thus the Bible is reinterpreted accordingly through that framework. Like other heresies, this one happens to also defeat salvation and the basis of Christianity:

The Gospel

Distorting Christianity to accommodate millions of years and Evolution is heretical and highly destructive to the whole basis for Christianity, because the chronological sequence of events in Genesis 1-3 are crucial to Christianity's cohesion. The Bible teaches that people can be reconciled into a right relationship with God (forgiven) because of the death of Jesus on a cross. In Christian Theology, this is made possible because Jesus is an 'atoning sacrifice' who pays for sin on our behalf. This is because the rightful consequence of rejecting/disobeying God is death (e.g. Gen 2:17, Rom 6:23), but if Jesus dies in our place, our punishment is taken upon himself, and we can be cleared of blame in God's eyes.

After Adam and Eve sinned, Genesis records that God placed 'a curse' on creation (Gen 3), and at this point God allowed bad things to happen in the world. However, if there was a large time period before man was created/evolved, this would mean there certainly was death, disease and destruction of various kinds before man (the effect of the curse). That is, bad things were created by God to be in the world before man rebelled from God. This would have a number of serious theological implications for Christianity:

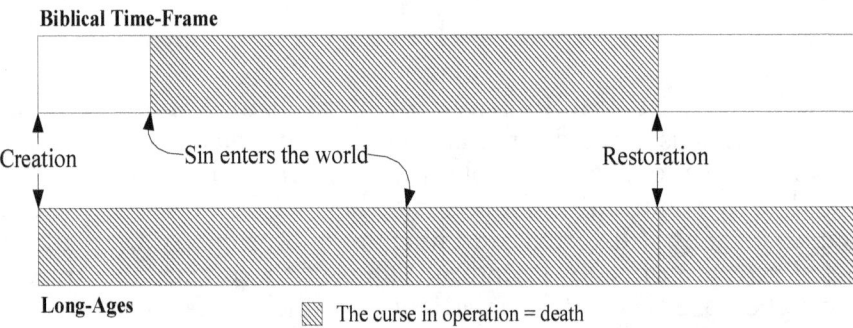

(1) If one accepts the Geological interpretation of billions of years, then in doing so we accept that there was death, disease, destruction and disasters long before Adam and Eve sinned. In the geological record we see billions of fossilized dead animals and people who were killed in the ancient world. There is also evidence of various types of diseases such as cancer and arthritis in the fossil record. We obviously see thorns and thistles and evidence of animals eating one another, and so therefore, by this chronology 'the curse' was operating within creation before Adam and Eve. This would imply that God intended such 'natural evils' to be a normal part of his "very good" original creation.

(2) In addition to being made clean and restored into a right relationship with God, the Bible promises a new creation where those who chose to live under God's loving care will forever be freed from all forms of evil and death. This is because the curse will be removed (Rev 22:3) and people will be restored into a perfect relationship with God *as they were at the beginning* before sin. Logically though, if death was always a part of the world before man's sin, then it would be hard to argue that there will be no death in heaven, because this would be something that God has originally created as good.

(3) Following the logic then, if death was present before sin, then sin is not the cause of death. This means that if death is not the consequence of sin, then sin cannot be paid for by death. Therefore, Jesus' death could not pay for our sin, because if death is not a consequence of sin, then his death, in substitute for our own, would have no effect on the debt we owe. Thus, if

millions of years of death before Adam and Eve was true, then there can be no forgiveness of sin by Jesus' death.

(4) Jesus is able to offer people life because he was able to overcome death (evidenced by the resurrection) through living a sinless life. Since he never sinned, the consequence of his life was eternal life. Thus, Jesus offers people life because he himself attained it through a perfect relationship with God. If, however, one accepts that death was present before sin, then to be consistent, there is no reason for a perfect life to be any different from our own.

Event	Biblical Time Frame	Long-Ages
Creation	No sin, no death	Death etc
Fall	Sin >> + Death	Death + Sin
Cross - Atonement	-Our Sin << His Death	Our Sin + His Death
Resurrection	No Sin >> No Death	???
Heaven	No evil >> no curse	???

Millions of years of Earth's history pre-dating sin would have the implication that there must have been various forms of evil and death before man. The implications of this are that God created these evils, heaven will contain such evils, Jesus' death cannot pay for sin (hence no forgiveness), and that Jesus cannot offer eternal life. Therefore, if we are going to be honest and consistent, there would be nothing in the Bible worth believing if man was not created 'at the beginning of creation' (Mk 10:6). We may as well throw the Bible away if the way that things are today are the way they have always been. Thus, it is not overstating the issue to see the acceptance of evolutionary history as a serious heresy defeating Christianity at its foundations.

If someone today was to teach other Christians that Jesus wasn't God, then they would label this person a heretic and remove them from any position of responsibility. Now, this person might argue "I'm a Christian. I have a personal relationship with God, I believe in his word and love other people. What's the problem?" And of course responsible Christians would say "Yes, but if what you are saying is true, then Jesus cannot forgive sins, none of us can be saved, and those other things would be meaningless." Well, it is exactly the same problem with millions of years pre-dating man! People can claim to be good Bible-believing, God-loving, honest Christians, but if what they are saying about millions of years was true, then no one could be saved and Christianity would be worthless.

This does not mean that such a person is not a Christian, but that they have an problem with how they define Christianity. At the same time as claiming to be Christian, such a person attacks the Christian worldview on which Christianity is based. Thus, the gospel that they believe in, which under-girds their faith, is a mismatch of different conflicting concepts producing a warped definition of Christianity. We should understand that to reject what Jesus teaches in the Bible (under the guise of 'interpretation') is an act of unbelief and sin, and is something that should have been repented of in the process of becoming a Christian. After all, to be a Christian one has to follow Christ and hold his opinion over their own. When one can be a 'Christian' and retain their right to choose what they will believe, it shows that we have a very flexible definition of a 'Christian' which is not in keeping with the Bible. However, since we can be certain of the historical basis of Christianity, there is no need to fashion such a compromised faith in the first place.

The Basis of 'Faith'

'Faith', by the biblical understanding (say Heb 11:1 for example), is about having confidence in what we *know to be true,* and therefore no one can have faith without knowledge of the truth.[185] One's faith is also limited by the amount of truth one is willing to accept, and strengthened by how certain someone is that what they believe is true. The Christian faith is not 'faith in faith', but faith in God. It is therefore generated by one's assurance that the *object of one's faith is true* and can be trusted. Thus, the basis of coming to accept that God exists and can be trusted (Heb 11:6) is dependent on what we understand to be the reality of the world we are living in, why we are in it, and how God fits into that. It is easy for a person to believe that God exists when they understand how the world requires a creative intelligence. It is just another step to accept the Christian faith when one looks at the historical basis of Christianity, where God has reportedly intervened throughout history and especially in the life, death, and resurrection of Jesus Christ. Thus, 'faith' is not based on our experience, or blindly based on what we hope to be true, but rather the Christian faith is based on what *is* true and therefore what *can* be experienced.

The foundational truths of Christianity necessary for faith are based on *actual events* in history. For example, the creation of the world, the creation of man, the place of God in our lives, the origin of deception, the loss of right relationship with God, the giving of a moral law, the word of God through prophets, the life and teachings of Christ, the death of Jesus on the cross, the resurrection, and the changed lives of Jesus' disciples because of the Holy Spirit, for example, are all historically known events of God's intervention in our world. We see that a biblical faith is always based on truth about the world: Throughout the book of Acts (e.g. Acts 2,3,10) the way the Apostles communicate Christianity is through grounding it in the events of the life and resurrection of Jesus. In 1 Corinthians 15:1-8, for example, Paul wishes to remind them of 'the gospel' he has preached to

---

185 See also Bible - Rom 10:14

them, and then proceeds to give them a list of historical events! Likewise Jesus gave people evidence in order to grow their faith. For example, he helped someone overcome their unbelief by performing a miracle (Mk 9:24). When John the baptist asked if Jesus was the Christ, he cited the physical evidence of his ministry (Lk 7:22). Also when Thomas doubted Jesus, he gave him the chance to see and touch Jesus for himself (Jn 20:26). The writer of John then says there were plenty more evidences they could have given, but that "these are written that you may believe that Jesus is the Christ, the Son of God, and that by believing you may have life in his name." (Jn 20:31) Peter and John say how they "cannot help speaking about what we have seen and heard" (Acts 4:20), and John also says that they proclaim what they have heard, seen and touched (1 Jn 1:1). Thus, it is unbiblical to suggest that the Christian faith can be removed from the facts of the world and events of history. Jesus himself also suggested that if people do not accept teaching on 'earthly' things that you can see and know, then surely they will not accept teaching on spiritual things (Jn 3:12).

Christianity, then, is based on truth and reason. The reason why someone chooses to follow Christ is entirely dependent on the sort of world that they believe they are living in. There would be no real reason to become a Christian if we do not live in a Christian world. For example, if God did not create us, and our separation from him does not explain the evil and suffering in the world, then the rest of Christianity that proposes to deal with that would be meaningless. What hope would Christ offer to restore/save people if the cause of man's problems were not related to him? The gospel, or essential/foundational truths of biblical Christianity are trustworthy because they relate to things we can know about the real world. Evolution, and millions of years thinking, directly attacks Christianity because it presents an alternative history, which if true, would refute the Christian gospel.

Distorting 'Faith'

Evolution and Christianity cannot both be true, because they make conflicting historical claims. However, many people who wish to avoid conflict choose to regard this as a non-contentious issue, and allow Evolution and Christianity to co-exist alongside one another. This may seem like an amicable arrangement, but the reality of the situation is that the 'facts' *are* in dispute, and choosing to avoid the conflict usually results in people drafting 'Science' and 'religion' into different realms of knowledge. Our culture often sees 'Science' as authoritative truth, whereas 'religion' is now seen as merely about a moral belief system. Christianity is culturally perceived as not attempting to explain reality and therefore not providing universal answers to life. However, if faith is ultimately based on evidence, to arrive at the position where 'faith' and 'facts' do not conflict with one another is to redefine 'faith' to be something which is non-factual (non-biblical). This would be to deviate from the biblical formula of a right faith in God, into a culturally-based pseudo-Christian religion. For example, many people will say that Evolution "does not affect their faith". However, if someone's faith was grounded in the gospel (the key truths about reality), then it would be nonsensical for them to claim that the actual truth of reality does not affect their understanding of the key truths about reality! This, in fact, is an admission that they have already bought into a form of Christianity that has no historical basis.

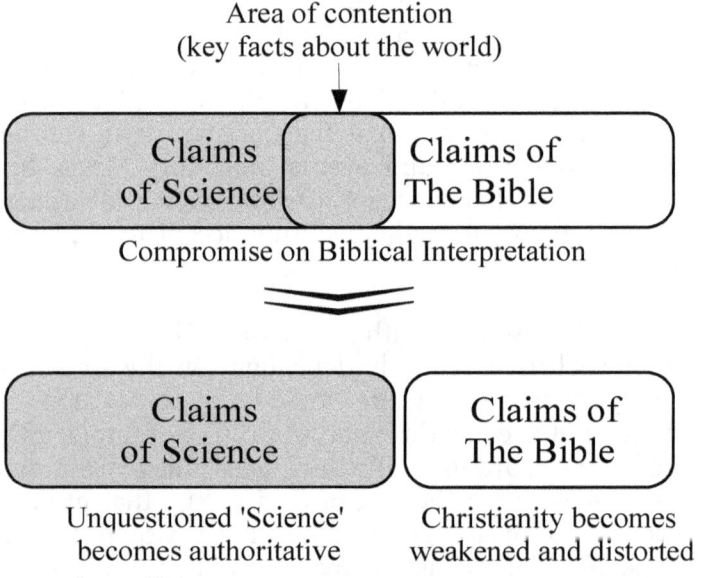

Thus, avoiding conflict about historical details inextricably leads to a redefinition of Christianity, and a faith not based on the gospel (key truths about reality). The Christian faith was meant to be based on something which was reliable and unchanging (the gospel), and if we don't believe in Jesus because of the facts, then the result is an irrational, subjective faith. Someone's faith is only ever as strong as the reason *why* they believe in Jesus. If these reasons lack an objective foundation, then that person's faith is always going to be disadvantaged and subject to cultural distortions.

Simply put, if Evolution was true then the Christian gospel would not be. Redefining this gospel so as to not be affected by the facts of reality, defeats the purpose of having a gospel. The result is that people's faith becomes increasingly experience-based, with little or no reliability[186] or transferability.[187] The perception that Evolution is true vilifies Christianity and makes it very difficult for anyone to accept as true. The perception that Christianity is irrational and based only on personal experience creates a perception that it is not universally true, and therefore does not

---

186 Many people begin to question their faith when they are challenged by Evolution.
187 It is not relevant to others who do not share that same experience.

apply to every person. Much of the cultural downfall in the perception of Christianity is caused by the fact that it has not survived in a battle of the fittest with Evolution. Be that as it may, we have now seen there is no reason to hold such positions, because Evolution can easily be challenged and refuted. Evolution is certainly no threat to the creationist position, rather it is a philosophical fall-back for people who do not *want* to believe in God. Furthermore, the gospel, with its basis in Christian Theism, is supported by a wealth of rational and historical evidence – only some of which is presented here.

Summary & Conclusion

Genesis forms an essential foundation to the rest of the Bible, and in particular, an essential basis for accepting the Christian Worldview.[188] If Genesis is not a historical narrative, then there is simply no basis whatsoever to believe that we live in a Christian world. An internal investigation of the Bible shows us that the only approach to Genesis that is consistent with the rest of scripture is to see it as a historical narrative. The chapters preceding this one have built up a case showing there is strong scientific/rational support for Genesis as being the true history of the World. Therefore, there are no valid reasons remaining for why someone could not accept that a straightforward reading of scripture aptly describes reality.

It is likely that people only reject a straightforward reading of Genesis because:

(1) They are reinterpreting the Bible in order to accommodate what they have heard *some* scientists say is true history. This is, in fact, taking those people's opinions to be more important than what God has said in the Bible. This accommodation of millions of years and Evolution has the effect of putting 'the curse' involving death and suffering before man ever arrived on the scene and sinned. As described, this defeats the purpose of Christianity completely, because there can be no redemption in Jesus Christ if death is not a consequence of sin. It is therefore heresy, and indicates a fundamental flaw in that person's faith.

(2) Or, a more popular approach today is to avoid the conflict of opposing truth claims (between biblical Creationism and Evolution). However, by doing this we inadvertently are forced to redefine 'Faith' and 'Science' so as not to overlap in the facts they claim. Christianity then becomes something that has no reliable grounding in reality, and therefore 'faith' becomes something that is irrational - having not been based on the gospel (which includes many facts about reality). This sets people up to fail, because the object of their faith subjective and unreliable.

---

188 The truth relating to our present reality which means Jesus is relevant to every person on the planet.

If Christianity has no factual basis, then it is merely a subjective religion, and therefore it lacks any degree of authority. At the other hand, if Creationism is correct, then Christianity is well anchored in Christian Theism, and therefore Evolution and millions of years of time is false. In reality, both viewpoints of history cannot both be right, and the implications of Creationism/Evolution affect everyone regardless of whether they are Christian or Atheistic. The Christian Worldview at least claims to be the reality in which we live - because it is based in creation. Someone who becomes a Christian ought to do so based on the understanding that the truth is we live in a Christian world (that Christ created), and therefore it is inescapable that to respond to Christ is the right thing to do.

Summary & Conclusions

Summary

(Ch.1) **Science and Faith** - Science and the Christian faith are not in conflict. In order to have the scientific framework we must make a number of assumptions about the reality we live in, and these assumptions are based on Theistic Worldview. That is, cause and effect is always true (leading to a supernatural cause of the Universe), we must assume a rationally ordered Universe (non-random), we are immaterial beings who can observe the physical world and understand its rationality, and that we are capable of perceiving the *right way* of connecting facts in what we call *logic*. All of these assumptions would only be true if we lived in a created world, designed by an entity that is supernatural, ordered, logical and infinitely intelligent. Therefore, there can in theory be no logical/scientific arguments against these founding assumptions, and therefore *true* science and Christianity *can never* be in conflict. Science favours the Christian Worldview.

(Ch.2) **Biological Evolution** - Macro-evolution is the concept that one particular broad group of animals can transform over time into another (for instance dogs to cats). There are only about five major arguments used to support this concept - despite the volumes of information committed to the cause. These arguments when identified and analysed do not support Evolution by descent with modification, rather, they provide support for biblical Creationism. The confidence with which some of this evidence is projected to the public is therefore misleading. For instance, the fossils of so-called ape-men do little to enhance the case for Evolution, most of them having been discredited a long time ago. Having apes so closely related to humans in fact provides a rational basis to argue that humans are uniquely created specimens, because the tiny *physical* differences fail to account for the *observed* actual differences between people and apes. Science favours Biological Design.

(Ch.3) **Abiogenesis** - Abiogenesis is the concept that life can (or at least did) come about through the random combination of chemicals. When examined, it is found that life is exceedingly complex, and the problem of creating life apart from God is insurmountable, bordering on ridiculous. The problem is a two-fold one: of obtaining the incredibly sophisticated array of chemicals and macromolecular machines (the hardware problem), and also obtaining the DNA sequence from which all life derives (the software problem). Correctly obtaining a functioning DNA sequence by chance has a prohibitively small likelihood, and ought not to be considered possible. Additionally, life is based entirely on 'information', which, by its very nature, necessitates an information-rich (intelligent) source. Science favours an Intelligent Design.

(Ch.4) **Cosmology** - The Big Bang is a Naturalistic model to explain the origin of the Universe, and it has been shown to be in violation of a number of simple scientific laws, and as such, is simply not a valid theory. Any Naturalistic explanation of the cosmos will inevitably fall short, because it does not explain how something came from nothing, what caused the Universe to begin, and the apparent organization of the fundamentals of physics - to name a few. There is also no explanation for how such an unlikely series of parameters coincided to result in our life-sustaining planet. A different cosmological model more in keeping with the facts (White Hole Cosmology) can be constructed, and this is consistent with a biblical view of God and history. Science favours a Creation Cosmology.

(Ch.5) **The Age of the Earth** – The basis of the millions of years' paradigm is an assumption about the past called Uniformitarianism. This assumption was formulated as an alternative to Flood Geology or Catastrophism, and has gained widespread popularity, despite the fact that it fails to explain many obvious geological phenomena. The evidence usually used to support millions of years of geological time only holds true if the assumption itself is true, and hence this circular reasoning does not support the millions of years paradigm. For example, radioisotope dating relies on Uniformitarianism, and therefore does not support millions of years without firstly assuming it! Catastrophism, on the other hand, is a more effective explanation for global geological phenomena, and is consistent with most of today's observable processes – which effectively limit the age of the Earth to thousands of years. Furthermore, if we do think through the biblical scenario of Noah's flood in scientific terms, we arrive at a history that accounts for what we see today, including how we can explain natural disasters in a world that God created to be good, and explains what happened to the dinosaurs. Science favours biblical History.

(Ch.6) **Biblical History** - Whether people acknowledge it or not, because the Bible claims to have recorded history "from the beginning", it does in fact affirm a recent creation (about 6000 years ago) in 6 ordinary days. (This is coupled with events such as Noah's flood and the tower of Babel, which explain many features of Earth's history that Evolution does not) Other views of Genesis simply contradict the rest of the Bible, and usually invoke embracing a view where death is chronologically placed before sin. This view obliterates the Christian message of salvation through Christ (gospel). Alternatively, many people avoid the argument with Evolution etc, by putting 'facts' and 'religion' into different, non-overlapping spheres of knowledge. This unquestionably results in a reconstruction of Christianity to be non-factual – a position that greatly detracts from the biblical narrative and therefore the basis of Christianity. Nevertheless, as we have previously seen, this is because of the 'creation secret' and the strident claims of evolutionists. Such positions are a result of being deceived into believing Evolution is a valid theory, when in reality factual support clearly favours biblical history. The Bible supports Young-Earth Creationism (biblical Creationism).

Overall

In each chapter of this book, we have started by looking at the key arguments used to support Naturalism/Evolution, and in each case, it has been shown that the so-called 'facts' used to support such views are nothing of the sort. It even often seems to be the case that these arguments better serve the creationist position when cross-examined! By contrast, the Creationist position involves a wider view of the facts, and presents itself as the most logical explanation - even though it is perhaps the most challenging one. It appears as though Naturalism/Evolution has been promoted as a scientific case, when in reality it involves sifting through the information, repressing that which supports Creation, and misleading people on the validity of whatever remains. It appears as though people are swept up into believing Evolution, not because of its factual validity, but because of the shameless filtering and distortion of information by Evolutionists to support only their preconceived philosophical view. If Evolution was an honest appraisal of the facts, then there would be no material to fill this book. Now, having revealed that there is a wealth of commonly concealed information supporting Creation (the creation secret), it becomes clear that Evolution has only succeeded as a way of thinking because the actual evidence against is has been kept a secret.

Naturalism/Evolution therefore does not qualify as scientific, because it is relentlessly a philosophical or religious viewpoint that attempts (unsuccessfully) to justify itself though its assertions about the facts, whilst at the same time suppressing those facts which lead towards supporting Creationism. Evolutionists are guilty of promoting their Naturalistic philosophy as truth, when this always involves misleading the public, suppressing information (that supports Creationism), and at times, teaching information that is known to be false. For example, evolutionists fail to mention that macro-evolution violates the laws of Genetics, Abiogenesis violates the Law of Biogenesis, the Big-Bang violates Newton's laws of external cause and effect, and Uniformitarianism says that present processes explain all of Earth's geology, when *they know* there are many examples of structures that are *not in formation today* from current processes. These are just a few of many examples of Evolution violating the most fundamental scientific laws, whilst evolutionists mislead people by boldly asserting their position as valid.

Furthermore, the current situation where 'only certain (naturalistic) facts are acceptable' impedes the progress of Science, because it opposes potentially fruitful scientific thinking not in line with these religious presuppositions. For example, adherence to the Big Bang has prevented scientists from presenting alternative theories of universe formation which are more credible. Uniformitarianism has repressed Flood Geology, a new and underutilised science with unknown benefits (for example, perhaps we could learn how to cheaply manufacture oil in large amounts[189], or manufacture precious stones replicating flood conditions[190]). Evolutionary biologists have wasted countless resources trying to prove Evolution, with very little advancement or benefit. For instance, some so-called 'vestigial organs' (such as the appendix) which have a function have even been cut out of the human body because of evolutionary assumptions, rather than studying it and developing diet and medicines etc to keep it healthy. We have in recent times begun to discover the health benefits of bacteria and virus therapy. Perhaps if we assumed that bacteria and viruses were originally created for a good purpose (only a tiny percentage of viruses have harmful effects), we would better understand how to use them to their full potential to help control global climate and cure illness etc.[191] It is not responsible education to refuse to deal with certain facts about reality because they might lead people to believe in a creator. This close-minded anti-intellectualism limits the effectiveness of what man can achieve through Science.

Consequences of Naturalism

Moral Evil

Evolution has also popularised a number of moral evils which ought to be considered. Naturalism is a philosophy which asserts that nothing beyond the physical exists, and such a belief attacks the very definition of what it means to be human – reducing man to nothing more than an animal. Additionally, Naturalism also demands there are no objective moral laws nor absolute truth above the majority opinion, for such beliefs imply a higher authority than man. Consider, then, what the implications might be if man is promoted as only a collection of atoms and molecules with absolutely no governing influences other than instinct and self-preservation.

Viktor Frankl, a holocaust survivor, put it this way:

"If we present man with a concept of man which is not true, we may well

---

189 Oil can be made synthetically under heat and pressure in 15 mins in a process called thermal depolymerization. Replicating these flood conditions on a larger scale could provide an inexpensive source of energy, say for developing countries.

190 This has actually been done by a creationist. Len Cram manufactures opals by a process he designed to mimic flood conditions. See Andrew A. Snelling, *Creating opals - Opals in months— not millions of years!* Available on http://creation.com/creating-opals.

191 Jerry Bergman, *Did God make pathogenic viruses?* Technical Journal 13(1):115–125, April 1999. Available on http://creation.com/did-god-make-pathogenic-viruses. Viruses control the survivability of bacteria, that control the rest of the ecosystem. It seems that viruses are in fact essential to all life on Earth. Virus therapy has also proven to be effective at curing HIV, target, and cure genetic diseases, and infiltrate the brain to cure neurological disorders.

corrupt him. When we present him as an automation of reflexes, as a mind machine, as a bundle of instincts, as a pawn of drive and reactions, as a mere product of heredity and environment, we feed the nihilism to which man is, in any case, prone. I became acquainted with the last stage of corruption in my second concentration camp, Auschwitz. The gas chambers of Auschwitz were the ultimate consequence of the theory that man is nothing but the product of heredity and environment – or, as the Nazis liked to say, "of blood and soil." I am absolutely convinced that the gas chambers of Auschwitz, Treblinka, and Maidanek were ultimately prepared not in some ministry or other in Berlin, but rather at the desks and in the lecture halls of nihilistic scientists and philosophers."[192]

Viktor here acknowledges that man is prone towards Nihilism,[193] and the more that we affirm the nature of man according to Naturalism, the more we feed the lack of moral responsibility. His experience of the Nazi mentality is that the evil they perpetrated was bred out of *an education* that presented man as nothing more than an animal, and without accountability of any sort. Hitler (and others such as Stalin, Pol Pot and Mussolini) was known to be a firm believer in evolutionary philosophy,[194] and only in a climate where such ideas are accepted, can a society begin to also accept such injustices that would otherwise be identified as evil. Of course there were mass murderers before there was evolutionary thinking, but the difference is that where once we could condemn such actions on the basis that murder was *morally wrong*, in a climate where Evolution is accepted as true, murder is perfectly consistent with its axioms. For example, the mechanism of Evolution has been called *"Natural Selection or the Preservation of Favoured Races"*[195]. Given this basis, there is nothing *inconsistent* about Hitler's preservation of favoured races by selection (mass murder). The more that a society embraces and popularizes naturalistic principles, the more negative an influence we can expect it to have on the moral climate.

Someone could also argue that religion has been responsible for an inordinate amount of evil as well. Consider, however, that those who do evil are operating in contradiction with universal moral conscience and clear commandments in the Bible. That is, so-called Christians, for example, who propagate evil are *wrong* and acting in an anti-Christian way. With no such moral absolutes however, there is no basis whatsoever to say that murder is wrong. Whilst it is hypocritical for a Christian to promote evil, under Atheism there is no such thing as good and evil, and

---

192 Viktor Frankl, *The Doctor and the Soul: Introduction to Logotherapy* (New York: Knopf, 1982), page 21

193 A philosophy of general lawlessness, or rejection of any moral standards. It is asserted far less often than it is practised.

194 The link between Evolution and Nazism is well documented in Weikart, R., *From Darwin to Hitler* (Palgrave Macmillan, New York, 2004).

195 The subtitle to Darwin's book: *On the Origin of Species - By Means of Natural Selection or the Preservation of Favoured Races in the Struggle for Life.*

would be hypocritical for them to promote the idea that people have any inherent value or moral responsibility. If Christians who do wrong disprove Christianity, then atheists who do *right* disprove Atheism!

Common observations tell us that if Evolution were true, the consensus at least is that it would be morally wrong to *act* as if it were true. Human beings *do have* certain indelible rights. This is the basis for all society and judicial law. People must not be treated as nothing more than animals because that is morally wrong according to our legal system. Murder is wrong because it violates the value of human life, and gives the power to bring about death into the hands of people who cannot restore life. Such issues as abortion are only contentious because society has now lost its ability to define what it means to be a human being. The genocide of millions of future people is justified only because of the lowering of the definition of a 'person' to be nothing more than a collection of atoms. If Evolution were true, there certainly would be nothing wrong with abortion. On the other hand, if every person brought into existence is someone with an identity, an inherent value, and a future to be protected, then it is a very sad day for mankind when the slaying of an innocent baby is considered a 'medical procedure'.

Evolution attacks both the moral reality and the definition of what it means to be human. The naturalistic worldview it promotes does not allow for human beings to have any sort of moral responsibility to an objective standard of right and wrong, and as such, it promotes Nihilism. It also does not allow for man to have an identity beyond the physical brain (a self), and as such, it denigrates the value of human life. For such things as the soul (the actual identity of a person within a body) and moral law cannot exist in a purely material world.

Naturalism and Real Life

A lack of scientific support for Atheism should come as no surprise to many people, because the life we experience everyday bears no resemblance to the sort of world we would expect if Naturalism was true. In short, the human experience cannot be reconciled with the concept that only material (atoms and energy) things exist.

The first hint that naturalism is an inadequate description of reality is that no one lives as though it were true. Naturalism prescribes a reality in which we are defined only by the laws that govern the movements of atoms and energy. Even an Atheist however, lives as though they had 'human rights', and values things such as love and goodness, detesting injustice and evil. Although an evolutionist argues that we are no different from animals, they certainly don't want to be treated like one! People talk about there being no real absolute truths in life when we discuss philosophy, but think in absolute terms when it comes to education, bank details, politics and how much tax they owe. Things like war, torture and suffering are universally regarded as bad, whereas generosity, compassion and charity are good. We all act as though there were *absolute moral realities* by constructing clever political and legal systems, and we imprison people who violate these. People live as though books and movies have *value*, and that a *good* movie is better than a *bad* movie. Hence, we believe in immaterial qualities even if our philosophy doesn't allow for them.

The point is that these are all concepts for which there is no physical reality. For instance, there is no *measurable* difference between compassion and cruelty, injustice and righteousness, but we still consider these to be clearly and diametrically opposed. Therefore, in these cases, there are no properties of matter that account for what we all know is real. The human experience generates a number of serious questions against Naturalism, because our experience is *characterized* by things like 'relationships', which, whilst being real, have no physical basis, and if they did, they would lose all meaning! If we considered 'love', for instance, to be only caused by the particular mixture of chemicals in the brain, then loving someone would not be a *free choice* – having been pre-determined by the chemistry. If this was the case, then love would have no real significance or meaning. This of course is not the case. Love *does* have real meaning because people *do* have the choice between hating and caring – it is not predetermined by chemistry. Hence we are held accountable for poor moral decisions.[196]

If Naturalism was true then:

➢ Nothing beyond the physical would exist.

➢ Everything would follow cause and effect, hence there would be no such thing as freewill or rational thought. Brain processes would all be automated responses to stimulus/environment (like a computer processor).

➢ 'Personality' could only ever be the result of genetics and environment, and the only thing that makes us human is our ability to process sensory data more effectively than other animals (a larger brain capacity). There is no self or individual identity.

➢ Atoms cannot violate physical laws, and therefore there could be no such thing as 'wrong' either practically (no moral law) or in terms of ideas (no true/false). Therefore Creationism could not be 'wrong' and Evolution could not be 'true'!

---

196 Unlike animals. People do not condemn sharks for biting people, or jellyfish for stinging – because that is what they are designed to do. We do not perceive them as capable of choosing to go against their instinct to attack humans and defend themselves. Some man-eating sharks are protected by law, and while they are allowed to eat us, we are not allowed to eat them!

> People could not have certain indelible rights by way of being 'human', because there is no external/objective moral reality beyond us. One bunch of atoms (a person) is not worth more than another (a rock).

> Since choice and morality would be illusions, no one could say or do anything 'wrong', and human responsibility is nonsense. Whilst a majority can come to a consensus about moral norms, there is no higher reason or moral code that those appeal to. Society could define anything to be 'right' – even murder.

> 'Love' and 'hate' are just words that correspond to chemistry inside the brain, but have no difference in actual value.

> Relationships would have no real meaning - being based only in the mixture of chemicals in the brain – which changes according to 'mood'.

> Once the chemistry had run its course, there would be no life after death, and no ultimate justice or reward. Ultimately nothing to live or die for.

Some Evolutionists even embrace these implications: *"Modern science directly implies that the world is organized strictly in accordance with mechanistic principles. There are no purposive principles whatsoever in nature. There are no gods and no designing forces that are rationally detectable. ... Second, modern science directly implies that there are no inherent moral or ethical laws, no absolute guiding principles for human society. ... Third, human beings are marvellously complex machines. The individual human becomes an ethical person by means of two primary mechanisms: heredity and environmental influences. That is all there is. Fourth, we must conclude that when we die, we die and that is the end of us. ... Finally, free will as it is traditionally conceived – the freedom to make uncoerced and unpredictable choices among alternative possible courses of action – simply does not exist. ... There is no way that the evolutionary process as currently conceived can produce a being that is truly free to make choices."* [197]

As the evolutionist says, the evolutionary process cannot produce a being that is truly free, but of course *he* was free to make that statement. The writer had the choice to write about Evolution or Creation, or about sharks, or nothing at all. It is therefore self-contradictory to say that there is no such thing as freedom of choice, whilst at the same time exercising his own freedom to choose! He expects us to exercise our own freewill in choosing to read what he has written, and furthermore to *choose* whether or not to accept it as true. If there really was no free choice, then why

---

197 Cornell University Professor William Provine, as quoted in Phillip E. Johnson, *Darwin on Trial* (1993), pg 126-127.

inform us about his view as opposed to another? If there are no designing forces, then how do we explain the appearance of design all around us? Doesn't this constitute rational evidence *for* such designing forces? If the world was *organised* according to mechanistic principles, then wouldn't it be impossible to discover rationality in its construction? Of course this is essentially what Science is about. If there are no moral or ethical laws, then on what basis do we accept what he is saying as 'right', compared to Creation, which he would argue was 'wrong'? If there are no guiding principles for human society, then should we accept Science/rationalism as a guiding principle to finding the truth? Science is not an invention (else we could change its guiding principles), rather it is what we have discovered to be the 'right way' of investigating the World. If there are no guiding principles for society, then why do we need laws and a *justice* system to maintain society? If the human being is, as he says, only a product of heredity and environment, then there seems to be little point in using *reasoning* in attempting to change people. According to his thinking, how would education and *truth* change people?

To make sense, his own statements require that we, in fact, are beings who can engage in freethinking and reason, who can distinguish between right and wrong, true and false, and these are the very things he attempts to argue against! Thus, since the evolutionary process, by his own reasoning, cannot produce beings such as we are, then Evolution as currently conceived is falsified! His ideas stem from the belief in a purely material world and are consistent with Naturalism, but these ideas can easily be shown to misrepresent reality, and cannot be affirmed without making self-contradictory, nonsensical statements.

Common sense, however, tells us that we do not live in a world where nothing more than atoms exist. Naturalism with its evolutionary basis is simply at odds with what we know about people and the reality that we live in. People live as if there is an absolute right and wrong (e.g. the legal system), as if there is absolute truth (e.g. an education system), as if there is a difference between love and hate (e.g. welfare systems), as if relationships have real meaning (e.g. marriage), as if humans have inherent value (e.g. murder is wrong), and have certain indelible rights (e.g. human rights movements). Society generally accepts these axioms, but they would not exist if there was nothing beyond the physical elements. Human society is unequivocally based on the acceptance that people are more than animals, and that there is a higher order beyond our own opinions. This strongly suggests that the world we experience is consistent with that depicted in Genesis, where man is a spiritual and physical being created within the reality of a perfect being we perceive as the example by which humans ought to live.

People Are More Than Brains

Whilst our brains process information received from the world, it is the *mind* that chooses what to do with that information. Whilst the mind uses the brain as a processing unit, the mind is more than just the brain. This is, in fact, a provable hypothesis: Scientists have discovered that it is possible to electrically stimulate part of the brain to cause the *involuntary* movement of an arm, for example. When the patient was asked to control the arm, they struggle to hold it still using their other arm. This demonstrates that whilst one arm was under the control of the electrically stimulated *brain*, the other was under control of the *mind*.[198] Thus, this experiment demonstrates that the mind is something in addition to the physical brain. The best explanation for this, and a host of other human functions already mentioned, is that a person is an immaterial being that inhabits a physical body (including the brain) – as the Bible suggests. The interaction between material brain and this immaterial soul is what we understand the 'mind' to be. It is where our true self engages with the world we perceive through the senses.

What We Can Know About God Through Human Experience

Freewill is the ability to choose what information we want *our brains* to process, and this is the precursor to rationality. This human ability demonstrates that the human mind is more than a unit that processes physical information alone, because there is also the 'self' that directs that unit as to what to process. The human mind is also able to perceive the way that a perfect mind ought to think, and therefore we have the ability to self-critique our own thoughts. We call this Logic, and it leads us to assess ideas for their validity and merit, hence the ability to decide between true and false.

We also have the ability to choose how we will act. We are not merely subject to our instincts or sensory responses, but we can choose to behave/respond in this way or that. Whilst we are free to choose how to live, we also have the ability to perceive the moral value of our actions and weigh their merit. For example, we know the difference between care and cruelty to another person - even when it does not affect our own survivability. Additionally, we can also perceive how the perfect person ought to act, and therefore we can self-evaluate our own lifestyle. We call this moral reality the Moral Law, and this leads us to assess the value of our actions, hence giving us the ability to decide between right and wrong.

---

198 Research conducted by Wilder Penfield as presented by Lee Strobel, *The Case for a Creator* (2004), pg 249.

The Logic of Moral Law & God

1. Moral Law and Logic exist - These two premises are impossible to deny, because one would have to argue that there is no right and wrong, or no true and false, whilst at the same time not making any 'right' or 'true' statements!

2. Moral Law and Logic are external/objective - Logic and morality cannot refer to a property of ourselves, because we would not be able to evaluate ourselves by ourselves. We would also have control over what we called right, which would defeat the purpose (we could treat anybody as we pleases and call it right!).

3. Moral Law and Logic originates from a personal entity - Logic and morality cannot refer to a static standard of order outside ourself (like the laws of physics) because they concern how a person ought to think or act.

4. The origin of Moral Law and Logic is perfect - Such an entity would have to be perfectly and unyieldingly above any moral or logical flaw, because only in such instance would there be a standard by which others could grade moral/logical action. For example, a standard which was half good and half evil would not be able to judge between good and evil.

5. The perfect, personal, external origin of Moral Law and Logic is 'God'.

Therefore, our experience of life demonstrates there is a personal reality by which our lives can be assessed. This corresponds to the Christian concept of 'God'. The creation story in Genesis depicts the creator as the perfect standard of truth and morality under whose authority and care we were created. The Bible also claims that Jesus Christ is that personal reality behind truth and love (right moral law).

What We Can Know About God Scientifically

The entity of 'God' is a necessary precondition to a number of realities that we observe. A pre-condition is a prerequisite on which rests the existence of something else. By analogy, if we find an apple, we can be certain that somewhere there must exist (or have existed) an apple tree, because a healthy tree is a necessary pre-condition to the existence of fruit – even if you cannot directly observe the tree. The nature of the apple also says something about the nature of the tree from where it came. In the same way, the Universe that we have surrounding us necessitates a supreme being as a prerequisite, some of the nature of which/whom we can deduce from the nature of the Universe. A creator God as depicted in the biblical Worldview can be shown to be *necessary* to account for what we see today.

(1) Necessary Preconditions to the Universe - In order for the Universe to exist as it does there would need to have been a first cause which is:

☑ External – Nothing (made of matter) can be self-caused. Everything that happens has a corresponding external cause, and everything that has come into existence owes its existence to something outside of itself. The Universe must have an external cause.

☑ Omnipotent – All powerful. No new energy can be manufactured within the Universe (1st Law Thermodynamics) and all energy is winding down (2nd Law). The Universe requires an external cause that is greater than the sum of all matter and energy within in.

☑ Omnipresent – (Transcendent/immaterial/supernatural). Any cause of the Universe (matter and energy) could not be a subset of the Universe. Hence a creator is beyond the creation, by definition, and therefore is not subject to the limitations of matter and energy (is supernatural). If God exists outside of space and time (having created it), we can therefore also deduce that 'God' can be accessible from all points within the Universe at the same time (transcendence).

☑ Omniscient – All knowing. The Universe has been ordered intelligently right from the foundational laws which govern physics and chemistry etc, through to the complicated encoding used to produce life (DNA). The potential for all kinds of knowledge comes from what has been imprinted in creation, and everything that we know as human beings we have learned from that source. Therefore, this source must have been greater in intelligence than the sum of all human knowledge.

☑ Eternal – The ultimate creator cannot be dependent on another cause for existence, but must be eternally existing (uncaused/outside of time). Being uncaused, it is more likely, then, that creation owes its existence to an *act of the will* of a creator, rather than a mechanistic cause.

(2) Necessary Preconditions to Life - In order for life to look as it does, the origin of life needs to be:

☑ Alive – The creator cannot be a static force such as gravity, but must be creative and alive – most likely a conscious being.

☑ Super-Intelligent – Nature is brilliantly and intricately designed for survivability on every level, implying an infinite, intelligent source.

☑ Information rich – matter cannot generate information, by definition, and therefore the information necessary for the language of life (DNA) must have originated from another source outside all matter, which in itself contained that information.

☑ Conceptual planning – interdependent macromolecular systems, cycles, interdependent body parts and even ecosystems, demonstrate prior planning before manufacture, since all parts would need to be present for each part to have any function.

- ☑ Forward thinking – life has been planned for long-term survival (for example programmed genetic variability etc), which shows imagination of future circumstances before design.

- ☑ Artistic – life is not just functional, but often elegant and beautiful. Life is not just cleverly designed for survival, but also aesthetically pleasing and interesting.

- ☑ Intentional – life appears to have been intentionally designed to display a non-naturalistic message, and present itself as the work of one designer. Sheep and snakes could easily appear the work of different designers (good and evil respectively) but for the fact that their basic cellular architecture is identical. The differences and similarities appear to be intentionally in place to show this.

(3) Necessary Preconditions to Human Potential - In order to account for human potential, there must be a God who is:

- ☑ Spiritual – Humans have freewill or the ability to choose between options - despite what our instinct or conditioning might tell us. This power over the physical self demonstrates that the true self (where the will resides) is not physical. Therefore, there must be an immaterial/spiritual creator of the immaterial/spiritual self.

- ☑ Personal - 'People' express *who they are* through their speech and actions. 'We' control *our* bodies and brains as we choose according to our personality (the true self), which cannot be comprehensively defined in terms of those physical parameters. There must be a source of human personality which in itself is a personal being like us - a human father.

- ☑ Rational – Atoms do not have the ability to violate cause and effect. So if we are able to think rationally and freely and choose between alternatives, then it is obvious that we have an ability that supersedes what atoms alone can accomplish. It would be necessary for there to be a rational creator (thinking God) to account for the ability of people to be rational and possess intelligence.

- ☑ Moral – It is also impossible for atoms to do 'wrong', since they cannot disobey natural law (physics and chemistry etc). There can be no *physical* basis to right and wrong, only a *moral* one. Since right and wrong are undeniable realities (no one can say this idea is 'wrong'), then there must be a moral source that accounts for our ability to be sensitive to what is 'good'.

'God', then, is a necessary precondition to the existence of the Universe, to life, and to the nature of human beings. We can also deduce some of the nature of God from these creations, and therefore define 'God' to be the all-knowing, all-powerful, eternal, personal being who created man in his likeness (personal, rational beings etc[199]). Hence, it is necessary for there to be a creator God matching the biblical description to account for reality. As we have already seen, alternative explanations, such as Evolution, utterly fail to overturn this conclusion.

Jesus as God

By analysing the physical world, it can be shown logically that God must exist and that we are part of a created world. We have also seen that within this world there exist inherent standards on the way things ought to be. We call these Moral Law and Logic. These are things that we did not invent and cannot change, and as properties of human nature, we must assume that they originate from a perfect person. In other words, they are realities of living within a universe where an omnipresent perfect being exists. Thus, although we are truly free to choose whatever we wish, our lives can be critiqued according to the way we ought to live in this world. There is an absolute basis for right and wrong, and it applies to all of us within the creation.[200]

One final question to consider is, given that we can find out something of the nature of God (what sort of being 'God' is) from the nature of creation, is there any basis for obtaining an understanding the *character* of God (*who* God is)? It has already been suggested that the Bible in fact provides accurate history when it comes to understanding the world that we live in. This is based mostly on Genesis of the Old Testament, but the same could also be said of New Testament (NT) history. There are plenty of resources dealing with this topic[201], but suffice to say that NT history is one of the most critiqued *and confirmed* books in history. The historical books of the life of Jesus (gospels) and the lives of the Apostles are written in such a way that they can be investigated and confirmed/refuted as accurate history. For instance, there are many historical details that allow them to be correlated with other historical works and archaeological finds, and where this has been possible, the Bible has been shown to be perfectly accurate.

NT history, then, records the life and teachings of Christ in detail, and is available for anyone to investigate. Christ's central claim was that in following him we would be living in right relationship with God because he

---

199 There are a large number of characteristics of people, suggesting that the source of that spirituality is like us, but perfect. It makes sense that the creation is consistent with the creator. The Bible suggests that the nature of man originated from the nature of God. It regards man as a special creation, unique in being like God and therefore able to relate to God.

200 If you disagree with this sentence, then you are suggesting it is wrong, and hence you prove that there is a right and wrong to live up to.

201 For example Josh MacDowel, *The New Evidence that Demands a Verdict*, and Norman Geisler, *Christian Apologetics*. See also Lee Strobel, *The Case for Christ*.

was God.[202] If this was true, then in claiming to be God, Jesus *claimed* to be the creator of the Universe and the creator of man. Continuing the logic then, we not only owe our existence to him, but we can also know how to live our lives because we perceive the Moral and Logical laws from him. As already described, God is the perfect moral being that allows us to distinguish between good and evil in thought and action, and if Jesus is God, he is that perfect person we are perceiving. Thus, universal human morality is Christian morality. Our sensitivity to truth and love is an indication of how sensitive our conscience is to the person of God himself. For example, when we experience love in its purest form, that is when we are the closest to God, because God is love (the perfect Moral law[203]). Jesus, in fact, claimed to be "the truth", meaning that rather than merely speaking the truth, he was the origin of truth itself (the perfect Logical law). Additionally, we can know God personally and become more sensitive to God because Jesus has told us how.

To be sure, Jesus' claims are outrageous. Jesus, however, did not make such bold claims without providing a rational basis on which to believe him. The basis for accepting Jesus Christ's creditability is again a case of historical inquiry. That is, there is ample evidence on which we can assess whether his claim to be God is logical or laughable. The New Testament bases belief in Christ on the events of his life: Jesus lived a sinless life and was perfectly loving to all people, especially those that society saw as unlovable. He lived a life characterized by miracles that showed he had authority over demons and nature. He showed the ability to heal all diseases and even raise people from the dead. He fulfilled literally hundreds of prophesies and allusions contained in the Jewish scriptures (Old Testament) predicting a coming messiah – the odds of which are astronomically small.[204] Finally, two days after he died he was transformed into a resurrected body and seen by over 500 people at one time. By this Jesus most convincingly demonstrated himself to be God by having power over death.

If God had chosen to reveal himself to us, it would be hard to imagine it being any more obvious than the combined revelation presented in the Bible, and in the nature of the person of Jesus Christ. This, together with the testimony of Jesus' followers alive today and throughout history, gives all people an objective basis by which God can be known personally.

---

202 Jesus Christ was crucified because he claimed to be God – a crime of blasphemy in that culture.

203 It is generally accepted that to love people unconditionally is the highest moral value that a person can exhibit. Jesus demonstrated this and the Bible teaches this.

204 The chances of anyone in history fulfilling just 8 of these prophesies that were totally outside of his control have been calculated to be $1/10^{17}$. The chances of anyone in history fulfilling 48 of the most well-known prophesies Jesus fulfilled is $1/10^{157}$. See Josh McDowell, *The New Evidence That Demands a Verdict* chapter 8 (1999), and Peter Stoner, *Science Speaks* (1963) Chicago: Moody Press.

The truth is no secret when people find out the information for themselves. This book has over-viewed and summarized some of the wealth of information that is readily available to anyone looking for it. Some key resources are listed below:

Websites

The key website that I have used to source material is:

Creation Ministries International: www.creation.com
Often I have referred to articles by their author and name followed by @ creation.com. In this case the article in question is easily found by typing the name of the article into the search engine there.

*This organization is well recognized and is very reliable in the information that they present. It is always respectfully communicated and well researched. They have built up such an enormous resource here that the Creation/Evolution information can be exhaustively sourced from here, and their affiliates.

See also:

The Institute for Creation Research: www.irc.org
Answer In Genesis www.answersingenesis.org
Creation Research www.creationresearch.org

Books:

References and quotes throughout this book have referred to:

Available from Creation Ministries International:
1. *The Creation Magazine* – Quarterly Magazine which is an excellent resource. It contains relevant news and articles etc written at a non-technical level.
2. *The Creation Answers Book* – Don Batten, David Catchpoole, Jonathan Safati & Carl Weiland available online at www.creationon.com/content/view/4018/
3. *The Answers Book* – Don Batten, Ken Ham, Jonathan Safati & Carl Weiland (1999), by Answers In Genesis Ministries (now Creation Ministries International).
4. *Refuting Evolution* – Answers In Genesis (Jonathan Sarfati) (1999).
5. *Refuting Evolution 2* - Answers In Genesis (Jonathan Sarfati) (2002), Master Books.
6. *The Biotic Message* – Walter James ReMine (1993), St Paul Science

Inc, Saint Paul, Minnesota.

7. *Refuting Compromise* – Jonathan Sarfati (2004), Master Books.
8. *In Six Days* – John F. Aston (1999), New Holland Publishers. This book is a compilation of essays by 50 scientists.
9. *The Lie* – Ken Ham (1987), Master Books, 18th printing (1999).
10. *Thousands Not Billions* – Don DeYoung (2005), Master Books.
11. *Dismantling the Big Bang* – Alexander Williams & John Hartnett (2005), Master Books.
12. *Starlight and Time* – Russel Humphreys (1994), Master Books, 4th printing (1997).
13. *The Genesis Flood* – John C. Whitcomb & Henry Morris (1961), Presbyterian and Reformed Publishing Company, 36th printing (1992).
14. *Grand Canyon* – Institute for Creation Research, edited by Steven A. Austin, (1994).
15. *In the Beginning Was Information* – Werner Gitt (1994), English Edition (1997), CLV.
16. *Creation and Time* – Mark Van Bebber & Paul S. Taylor, 2nd ed (1994), Eden Communications.
17. *15 Reasons to Take Genesis as History* – Don Batten & Jonathan Sarfati (2006), Creation Ministries International.
18. *Darwin on Trial* – Phillip E. Johnson (1991), Intervarsity Press, Illonois, 2nd Ed (1993).

Other Key Books I have used:

1. *The Case for Faith* – Lee Strobel (2000), Zondervan Publishing House, Grand Rapids, Michigan.
2. *The Case for a Creator* – Lee Strobel (2004), Zondervan Publishing House, Grand Rapids, Michigan.
3. *Christian Apologetics* – Norman Geisler, Baker Books (1976).
4. *The New Evidence That Demands A Verdict* – Josh MacDowell, Thomas Nelson Publishers (1999).

Special Thanks

*Special thanks to visiting speakers of Answers In Genesis/Creation Ministries International, the writers of their resources (internet articles, books, and in particular the Creation Magazine), and the CMI webpage resource, which over the years have educated me, encouraged me, and provided specialised resources on innumerable topics.

About the Author

Paul Garbett is a New Zealand based author who develops resources to help people to find out about God, grow in their Christian faith, and to grow healthier Christian communities. He specialises in apologetics, mission, personal growth, relationships and church structure.

To contact Paul, access more of his resources or find out about other books, visit his website at:
www.faithforward.weebly.com

www.ingramcontent.com/pod-product-compliance
Lightning Source LLC
Chambersburg PA
CBHW051315170526
45166CB00002B/558